THE COMPLETENESS OF SCIENTIFIC THEORIES

THE UNIVERSITY OF WESTERN ONTARIO SERIES IN PHILOSOPHY OF SCIENCE

A SERIES OF BOOKS
IN PHILOSOPHY OF SCIENCE, METHODOLOGY, EPISTEMOLOGY,
LOGIC, HISTORY OF SCIENCE, AND RELATED FIELDS

VOLUME 53

MARTIN CARRIER

Center for Philosophy of Science, University of Konstanz, Germany

THE COMPLETENESS OF SCIENTIFIC THEORIES

On the Derivation of Empirical Indicators within a Theoretical Framework: The Case of Physical Geometry

SPRINGER SCIENCE+BUSINESS MEDIA, B.V.

Library of Congress Cataloging-in-Publication Data

Carrier, Martin.
The completeness of scientific theories on the derivation of empirical indicators within a theoretical framework the case of physical geometry / by Martin Carrier.
p. cm. -- (The University of Western Ontario series in the philosophy of science ; v. 53)
Includes bibliographical references and index.
ISBN 978-0-7923-2475-1 ISBN 978-94-011-0910-9 (eBook)
DOI 10.1007/978-94-011-0910-9
1. Science--Philosophy. 2. Science--Methodology. 3. Physical measurements. 4. Geometry. I. Title. II. Series.
Q175.C335 1993
530.1'6--dc20 93-30527

ISBN 978-0-7923-2475-1

printed on acid-free paper

To Gabi
With Love

EDITOR'S PREFACE

Earlier in this century, many philosophers of science (for example, Rudolf Carnap) drew a fairly sharp distinction between theory and observation, between theoretical terms like 'mass' and 'electron', and observation terms like 'measures three meters in length' and 'is −2° Celsius'. By simply looking at our instruments we can ascertain what numbers our measurements yield. Creatures like mass are different: we determine mass by calculation; we never directly observe a mass. Nor an electron: this term is introduced in order to explain what we observe. This (once standard) distinction between theory and observation was eventually found to be wanting. First, if the distinction holds, it is difficult to see what can characterize *the relationship* between theory and observation. How can theoretical terms explain that which is itself in no way theorized? The second point leads out of the first: are not the instruments that provide us with observational material *themselves creatures of theory*? Is it really possible to have an observation language that is entirely barren of theory?

The theory-ladenness of observation languages is now an accepted feature of the logic of science. Many regard such dependence of observation on theory as a virtue. If our instruments of observation do not derive their meaning from theories, whence comes that meaning? Surely – *in science* – we have nothing else but theories to tell us what to try to observe.

It is Martin Carrier's contention in this book that there exist theories that function as their own observation languages. Such theories are capable, without the intrusion of non-theoretical elements, of analyzing their own methods of measurement. Carrier calls this feature of mature theories 'completeness', a term applied by Misner/Thorne/Wheeler to those theories capable of analyzing every experiment in their domains of application.

Carrier's important book develops and clarifies this idea of completeness as a special form of theory-ladenness. It then applies the idea of completeness to the case of physical geometry. Throughout the book, the philosophical analysis is preceded by the presentation of historical case studies of theories in science.

Dr. Carrier is a member of the Zentrum Philosophie und Wissenschafstheorie in Universität Konstanz, where he holds the rank of Akademischer Rat. He earned his Ph.D. at Universität Münster and completed his habilitation at Universität Konstanz. He has written a number of papers on scientific methodology, scientific realism, and the atomism of seventeenth- and eighteenth-century physics and chemistry. His ground-breaking work on the conceptual development of affinity theory in the eighteenth century has attracted an international audience. He is co-author (with Jürgen Mittelstraß) of *Geist, Gehirn, Verhalten. Das Leib–Seele-Problem und die Philosophie der Psychologie*, Berlin: de Gruyter (1989); revised English version, *Mind, Brain, Behavior. The Mind-Body Problem and the Philosophy of Psychology*, New York: de Gruyter (1991).

Robert E. Butts

PREFACE

Sometimes the need of a theory to resort to an auxiliary so as to achieve a desired effect has fateful consequences. This is evidenced in a particularly dramatic fashion by the sad story of the physicist Spalanzani, as recounted in the Second Act of Jacques Offenbach's *Les Contes d'Hoffmann.* Spalanzani had craftily assembled a mechanical doll which he hoped would pass as a living being. Alas, however, his skills in mechanics failed to accomplish working artificial eyes; eyes, that is, which were suited to deceive a future beholder about the mechanical nature of the doll.

Since mechanics had reached its limits at this juncture, Spalanzani felt compelled to turn to optics – embodied in the optician Coppelius – for some additional help. Coppelius indeed succeeded in manufacturing appropriate eyes. But in this enforced cooperation lay the seeds of the tragic development that was subsequently to unfold. Details aside, the two scientists quarreled about money with the eventual result that Coppelius dismantled the outcome of their joint venture.

The moral is that if mechanics had managed to achieve the purpose at hand relying on its own theoretical resources, the desaster would have been avoided. It was the need to draw on the results of some other theory that gave rise to the tragic finale. I conclude that theoretical autonomy is of some use from time to time.

Having thus made plausible the practical importance of conceptual autonomy, it is time to announce that this book is devoted to the study of precisely this subject matter. I'm concerned with the capacity of theories to analyze the measuring instruments attached to the theoretical states introduced within their framework. Such theories manage to cover the corresponding empirical ground on their own; that is, without having to rely on additional theoretical help. It appears immediately that Spalanzani would have benefitted from a theory of the kind. The conclusion to be drawn is that the subject matter addressed in the present study is of some real-life importance. In other words, I am tackling a philosophical problem with some import on existential aspects of human life.

The work on theoretical autonomy was not done in mental autonomy. So I'm grateful to a large number of persons who were kind enough to further this study. Gereon Wolters drew my attention to problems in the philosophical interpretation of the theory of relativity; the present book grew out of the occupation with these problems. Claus Lämmerzahl repeatedly gave me technical advice in matters of General Relativity. An earlier version of the text constituted my *Habilitationsschrift*; it was written in large part while I was a Visiting Fellow at the Center for Philosophy of Science of the University of Pittsburgh. I owe much to the Center's warm and inspiring atmosphere. Completion of the work was supported by a grant from the *Deutsche Forschungsgemeinschaft*.

The *Habilitationsschrift*-predecessor of the present book was read by Jürgen Audretsch, Jürgen Mittelstrass and Alexander Rüger. All of them came up with valuable suggestions for developing and improving the argument. Thus they essentially contributed to giving the text its present shape. Peter McLaughlin, finally, checked my English style and moreover urged me to further clarify various claims. I heartily thank all of them.

Martin Carrier — *University of Konstanz*
May 1992 — *Center for Philosophy of Science*

TABLE OF CONTENTS

PART B. THEORY AND EVIDENCE IN PHYSICAL GEOMETRY

INTRODUCTION

Science is often regarded as an experimental enterprise; it is characterized by its recourse to measurements. This is certainly true and important in addition. There is, however, a further aspect involved. Namely, we would hardly count blind trial-and-error procedures, whose application, after all, betrays an outstanding experimental leaning, as the high point of scientific investigation. Simply collecting and listing some easily accessible observations is generally not considered great science. And rightly so, for what is missing here is a theory guiding the experimental scrutiny.

Significant and seminal experiments rely on fruitful theories. Put the other way around, if an experiment is performed against an inadequate theoretical background its results may later easily turn out to be worthless. This is especially true in cases in which a theory fails to recognize a factor that is actually influential on the empirical outcome. A historical case in point is the devastating effect that Claude Berthollet's discovery of mass action around 1800 had on the significance of experiments carried out on the basis of 18th-century affinity theory. According to affinity theory, the combination of chemical substances is governed by affinity forces, i.e., short-range chemical forces that were assumed to be specific for each pair of substances and independent of other physical magnitudes such as temperature. The capacity of a substance to replace another substance in a compound was regarded as indicating empirically the relative strength of the affinity forces involved. If a substance C was able to supplant a constituent B in a compound AB (i.e., $AB + C \rightarrow AC + B$), this was taken to indicate that the affinity between A and C is stronger than between A and B. On the basis of such substitution reactions, affinity tables were established for every substance. These tables rank-ordered substances according to their capacity to replace constituents in compounds of a given substance. The above reaction would thus allow one to give the following midget affinity table for the substance A: C, B.

After two generations of chemists had been busy with ascertaining affinity tables, Berthollet came up with a theory according to which the reactions a given substance performs depend crucially on the quantity of the substance present. In particular, an alteration of the respective quantities may reverse the reaction altogether, with a concomitant reversal in the relative affinity strengths to be assigned. Since none of the experimentalists had bothered about quantities when examining combining properties, their results contained an uncontrolled factor that was of considerable influence on the outcome. After Berthollet's theory had been accepted, all the earlier experimental results thus appeared completely useless. As Dietrich Karsten, a chemist of the period, noted in deep regret or even despair:

> All our hitherto assumed laws of affinity and all affinity tables, which have been the fruit of many active years of the preceding century, no longer hold true. [...] We are left with the sad certainty that all the investigations hitherto carried out on that subject, and that all the efforts of many of the greatest and most distinguished chemists to promote this important part of chemistry, have been in vain (Karsten 1803, 136, translation mine; cf. also Carrier 1986a, 383–384).

If it turns out that a theory has failed to take into account a variable whose magnitude actually is of influence on the results, a large class of scrupulously registered empirical data may lose all scientific significance. Their only remaining use is to fill the waste bin – or to instruct historians and philosophers of science.

Closely related to the theoretical neglect of a substantial influence is another type of theoretical misinterpretation. Experimentalists may be misled by an inadequate theory in that they look into the wrong quantities. That is, they examine aspects of a phenomenon that are too complex or liable to additional distorting influences and thus do not exhibit any simple, theoretically significant regularities. Experimentalists may fail to unearth important lawful relations inherent in a phenomenon without a theory that specifies them beforehand. Take the case of Brownian motion as an example. As Robert Brown discovered in 1827, microscopically small particles suspended in a fluid (such as pollen of plants floating in water) carry out a kind of "dancing motion". That is, these tiny particles fluctuate in an apparently irregular fashion.

In the latter part of the 19th century the attempt was made to apply the kinetic theory of heat to this phenomenon. In light of that approach, Brownian motion was interpreted as the observable impact of the invisible thermal motions of the fluid molecules postulated by kinetic theory. These molecules were assumed to collide erratically with the suspended particles thus causing the latter's accidental fluctuations. The early kinetic treatment started off from the so-called equipartition theorem according to which the mean kinetic energy of all particles is equal in thermal equilibrium. Consequently, the attempt was made to measure these energy values, that is, the average velocities of the Brownian particles. As it turned out, however, the outcome of these experiments conflicted considerably with the theoretical expectations.

Only after the Einstein–Smoluchowski theory of Brownian motion had been formulated in 1905, did it become clear what the origin of this failure was. The theoretically misguided observers had aimed at unsuited quantities, namely, velocities. On the basis of the Einstein–Smoluchowski theory, the significant quantities are the mean displacement and the vertical height distribution of the Brownian particles. It is these quantities for which simple and meaningful relations could be derived from the kinetic approach. By contrast, as Einstein showed, considering velocities leads right into a deadlock. Because of the rapid viscous damping of the particle velocities and the particles' equally rapid acceleration through molecular impulses, the observed average velocity continually increases as the time interval of the observation is diminished. As a result of these distorting influences, the instantaneous velocities of the Brownian particles exhibit no detectable regularity; they simply behave in an erratic and unfathomable fashion (cf. Brush 1976, 656, 681–682). As Stephen Brush summarizes this episode, experimentalists

> were simply measuring the wrong thing until Einstein pointed out that only the ratio of mean-square displacement to time could be expected to have any theoretical significance. One can hardly find a better example in the history of science of the complete failure of experiment and observation, unguided (until 1905) by theory, to unearth the simple laws governing a phenomenon (Brush 1976, 682).

Sometimes we are in need of a theory to direct our attention to those quantities whose investigation is to prove fruitful.

These examples demonstrate that the quality of experiments is crucially dependent on the quality of the theories invoked for their set-up and their interpretation. Inadequate theories may effectively preclude or thwart significant experimental work. In the present study I wish to address one of the variegated ways in which theory may intrude into experiment and observation. Namely, a theory may be used as observation theory; that is, it may be used for analyzing the relevant experimental set-up, for associating theoretically significant quantities with directly observable ones, or for correcting the raw-data obtained. Observation theories are applied to establish a connection between immediately accessible data such as the readings of measuring instruments, on the one hand, and the evaluation of theoretically assumed entities, on the other. Observation theories thus license the inference from an empirical event to a theoretically described state of affairs.

Interestingly enough, there exist particular cases in which a theory figures as its own observation theory. In such a case, the theory is capable of analyzing on its own the measuring procedures associated with theoretical states specified within its framework. This peculiar and noteworthy feature I call completeness. This term is borrowed from Misner–Thorne–Wheeler who apply it to theories that are sufficient for analyzing every experiment in their domain of application (cf. Misner/Thorne/Wheeler 1973, 1067). As a matter of fact, the property of completeness had been alluded to some decades before Misner–Thorne–Wheeler coined an appropriate term for it by Albert Einstein and Herbert Feigl. For that reason I speak of *Einstein–Feigl completeness.* What I intend to do here is, first, to elaborate this concept and to make it precise and, second, to apply it to the case of physical geometry.

These two tasks are addressed in the first part and the second part, respectively. The first part is devoted to clarifying some aspects of the relation between theory and evidence in general, and to the question of a theoretically self-sufficient treatment of the relevant observational procedures in particular. The argumentation proceeds roughly along the following lines. The first chapter is introductory in nature. It outlines some accounts of the semantics of scientific concepts and explores three types of theory-ladenness. One of these types, namely, measuremental theory-ladenness, proves to be of particular interest for the purpose of the present investigation.

The second chapter delves into the issue of Einstein–Feigl completeness. This feature is introduced as a special form of measuremental theory-ladenness. The central problem emerging with respect to complete theories is that in testing them one may get stuck, at least on the face of it, in a circularity or an infinite regress. If a theory that is supposed to explain certain measurement results at the same time provides the theoretical apparatus necessary for obtaining these results in the first place, the risk of circularity is obvious. It is the examination of the possible difficulties involved in the test of complete theories that represents the major concern of this book. It turns out that there indeed exists a variety of circularity problems that go along with particular types of completeness. From this it follows that complete theories might suffer from testability problems. On the other hand, test strategies can be specified that generally allow these problems to be defused. The result is, accordingly, that apart from somewhat pathological instances, Einstein–Feigl completeness does not have a detrimental impact on the corresponding theory's testability.

An additional issue surfacing in this connection is the need sometimes to rectify distorted empirical data. If the measurement of a quantity is corrected relying on laws that are part of the same theory as the quantity we wish to evaluate, we are dealing with another aspect of completeness. Such correction procedures may suffer from a circularity that is of an even more serious nature than the one referred to before. It will turn out that these difficulties set some limitations on carrying out corrections in a complete fashion. The third chapter expands on some examples from physics and psychology. These examples are intended to bring out the basic idea of completeness in a more concrete and intuitive fashion.

The second part is concerned with the application of the concept of completeness to the case of physical geometry. The latter is related to the notion of completeness in a two-fold fashion which is addressed in chapters IV and V, respectively. Chapter IV deals with the thesis of the conventionality of geometry. The connection between this thesis and the completeness issue is that the test problems surfacing with respect to the latter have the same logical structure as those underlying Hans Reichenbach's arguments for that conventionality. Conventionality is thus a special instance of a failed attempt to accomplish completeness.

In chapter V standard General Relativity is discussed. The reason for picking General Relativity as my chief example is that, as far as I can see, it constitutes the most complete theory ever proposed in science. In particular, it is possible, while relying almost exclusively on the theoretical resources it provides, to construct and analyze instruments suitable for the measurement of the relevant spatiotemporal quantities. Moreover, it is likewise possible to specify on that basis a large number of distorting influences and to ensure their absence empirically. Thus, the attachment of empirical indicators to theoretical states is established and the inference from (possibly distorted) evidence to the corresponding theoretical states is licensed by General Relativity itself. And it turns out, finally, that the testability of the theory remains unimpaired by this distinction.

Chapter VI, finally, resumes the discussion of the conventionality thesis. This time, however, emphasis is placed on the examination of the viability of the thesis itself. The question is whether physical geometry is conventional in a philosophically interesting sense. The answer to this question will be affirmative. I claim, that is, that geometry is indeed conventional in important respects.

Emphasis on examples pervades the whole book. That is, I proceed inductively by first exploring particular cases and afterward capturing their import in more general terms. I derive the relevant concepts largely from considering concrete examples. In my view, philosophy of science should be done in close touch with actual scientific practice. The disadvantage that goes with that commitment is that the presentation is not as succinct as it otherwise could have been. Presenting a conception in an abstract and general way is usually more elegant than the pedestrian case-by-case procedure I apply. In particular, since one and the same conceptual trait is often discussed by way of different examples, the reader may feel haunted by a *déjà-vu* experience on some occasion or other. But this is the price one has to pay for making it plausible that the philosophical conception one wants to expound is applicable to scientific practice. This aim can obviously not be reached from a bird's eye view.

For this reason the discussion sometimes presupposes a certain scientific background on the part of the reader. On the other hand, the technical skills required are kept at a minimum; familiarity with high school physics should in general suffice. It is assumed, that is, that the reader has some knowledge of the basics of classical physics including Special Relativity. The fundamentals of General Relativity will be developed in an intuitive and non-technical fashion in the course of the book itself. The same holds analogously with respect to quantum mechanics and psychoanalysis which I also explore from the perspective of their possible completeness. In all these cases a sketch version of the respective theory's content is provided.

To repeat, going into the particulars of substantive theory is essential if one wants to show that the conception one is about to develop is, first, of a down-to-earth nature and, second, does not express a singular phenomenon but rather a trait pervading all science. And as I said before, these objectives can only be reached by the repeated presentation of the same feature in different contexts. In music, a sequence of notes becomes a tune only by repetition; typically, musical structure is created by recurrence. So, the reader is invited to follow the multifarious variation and development of the completeness theme.

CHAPTER I

THE THEORY-LADENNESS OF OBSERVATION AND MEASUREMENT

1. THE VERIFICATIONIST THEORY OF MEANING AND THE DOUBLE-LANGUAGE MODEL

§ 1. It has been part of the standard empiricist creed for centuries that the meaning of concepts and the meaning of statements likewise derive exclusively from their relation to experience. But only the Vienna Circle (and its intellectual subsidiaries) managed to sharpen the notion inherent in this contention. As regards the meaning of concepts, one of the relevant approaches is *operationalism*, developed by Percy Bridgman in the 1920's. According to operationalism, every scientific concept obtains its meaning by being tied to a measuring procedure. The claim is that a concept's meaning is indeed comprehensively characterized by that procedure. As Bridgman put it, "the concept is synonymous with the corresponding set of operations" (Bridgman 1927, 5).

This approach entails a one-one relation between concept and measuring procedure. That is, different concepts must be associated with different procedures, and different procedures always constitute different concepts. Measuring lengths in different ways, e.g., by arranging measuring rods or by recording travel times of light, generates different concepts of length, i.e., "rod length" and "optical length". If experience shows, then, that the values obtained by the two procedures approximately agree numerically within their common scope, then this fact is to be expressed as the empirical generalization: rod length and optical length roughly coincide (cf. Bridgman 1927, 10, 16–18).

The operationalist account of the meaning of concepts is closely related to the *verificationist account* of the meaning of statements. According to the latter account, the meaning of a proposition is determined by the way in which this proposition can be put to empirical scrutiny. In the words of Moritz Schlick: "The meaning of a proposition is the method of its verification" (Schlick 1936, 120). Meaning derives from the confirming evidence. More precisely, a statement or proposition is synonymous with the class of those possible empirical results which are suited to support it. If a statement is confirmed by some observational data *a*, *b*, *c*, then this statement refers to *a*, *b*, *c* and to nothing else. A statement is translatable into its observational consequences.

This implies in particular that sameness of observational consequences entails sameness of meaning. Take psychological states as an example. All statements that purportedly express inner states of other human beings (such as "John is happy") can only be confirmed by recourse to overt behavior. Consequently, such statements actually refer to behavior, and not to some hidden mental causes of this behavior. It makes no sense to say that behavior is produced by some mental state since the only way to ascertain the existence of the latter is by way of the former. No observable differences correspond to the conceptual distinction between mental causes and behavioral effects, and for that reason it is meaningless (cf. Carnap 1928, 316–321, 334–338). The verificationist theory of meaning thus implies that the reference of a statement is always identical to the evidence confirming it.

One of the difficulties this approach runs into is that it is at variance with some aspects of the linguistic practice of science. There exist scientific concepts to which observational data are attached in a very loose way. In view of concepts such as "electromagnetic potential" or "quantum mechanical state function", it seems odd to contend that all scientific concepts derive their meaning from the confirming evidence. Accordingly, mature Logical Empiricism attenuated the requirements imposed on meaningful concepts. The model that eventually emerged in the 1950's was the so-called *double-language model*; it was soon to become the "standard view".

§ 2. The double-language model distinguishes between two linguistic levels, the *observation language* and the *theoretical language*. The former is characterized by the condition that all its fundamental descriptive concepts refer to immediately observable objects or properties, and that all its derivative concepts can be explicitly defined on the basis of the fundamental ones (relying, in addition, on the rules of logic). Accordingly, every statement in the observation language is directly verifiable by experience. That is, we can decide by simple, unaided inspection whether a given observation statement is true or false. In the case of observational generalizations (such as "all ravens are black") repeated inspections may be called for in order to arrive at an appraisal, if tentative, of their credibility. But even in that case we need not resort to any theoretical means for examining their validity. Unaided observation is sufficient to confirm or disconfirm every observation statement.

It appears from this description that the properties attributed to the observation language virtually coincide with the features that were formerly supposed to characterize the language of science as a whole. In other words, the verificationist theory is still supposed to hold for the observation language; on the other hand, its validity is now restricted to this observation language. And in addition to it, a theoretical language is conceived that is distinct from the observation language. The terms of this theoretical language are introduced by the central postulates of some theory and refer to entities such as electrons or light waves that are not immediately observable. Theoretical terms are not explicitly definable by concepts from the observation language. Rather, the connection between the two levels is established by so-called correspondence rules. Correspondence rules contain at least one predicate from both linguistic levels and thus attach one or several empirical procedures to a theoretical term or to a function of such terms. The set of correspondence rules confers empirical import to the at first uninterpreted theoretical structure.

Hence, the correspondence rules do not necessarily establish one-one relations between theoretical and observational concepts (and this is one reason, among others, for the fact that the correspondence rules are not suited as definitions for the theoretical terms). Rather, a variety of empirical indicators may be associated with one and the same theoretical term; and conversely, it may be that only a logical function of several theoretical terms – but not each of them separately – is linked to some observable phenomenon. Such a set of correspondence rules endows a theoretical structure with a partial empirical interpretation. This interpretation is partial in that it, first, typically fails to associate an empirical counterpart to every theoretical term and is, second, subject to historical change. In the course of the development of a theory the relevant correspondence rules may be altered; some may be added and some deleted. A further characteristic of theoretical terms is that their relation to observation terms is typically brought about by means of natural laws. That is, the correspondence rules are not semantic stipulations but rely on laws instead.

These peculiarities of theoretical terms can be exemplified using the concept "temporal duration". Durations can be measured by various processes or devices. One may invoke the annual revolution of the earth around the sun; or one may apply pendulum clocks, quartz clocks, atomic clocks and the like. There are a multitude of indicators associated with this concept. In addition, the set of accepted indicators changes in time. Atomic clocks were added to the indicator set only within the last decades. Conversely, Galileo started his investigations on pendulum oscillations by relying on his pulse beat as a measure of time. And this process is certainly no longer regarded as a reliable indicator of duration. Similarly, the daily rotation of the earth was accepted as a time standard until it was recognized that this rotation was gradually slowing down as a result of tidal friction. In these cases a correspondence rule has been abandoned. Finally, it appears from these considerations – and will be elaborated more thoroughly in II.3 – that the suitability of an empirical phenomenon as an indication of a given theoretical state is judged in light of a theory. The use of a phenomenon for this purpose is typically based on a theoretical analysis of this phenomenon. We know that a quartz clock reliably measures duration because we know how a quartz clock operates. The relation between a theoretical state and its empirical indicator thus holds by virtue of physical laws.

§ 3. The reasons for admitting theoretical terms are primarily methodological in nature. Laws framed in theoretical language typically combine or unite various empirical generalizations. Theoretical laws thus conduce to an economy of expression and a systematization of content that is absent in a mere collection of empirical generalizations. It is noteworthy in this context that most of the advocates of the double-language model retained the view that experience is the only source of the meaning of scientific concepts. There is no "theoretical surplus meaning"; strictly speaking, the theory merely constitutes a formal algorithm or a set of mathematical relations. It is through the correspondence rules alone that meaning is conferred to the quantities contained in the algorithm. There is no meaning apart from and in addition to empirical interpretation.[1]

As regards the relation between theory and evidence, it is characteristic of the standard view that the line of demarcation separating observation language from theoretical language is drawn in a *universal* fashion. This is to be understood as follows. It is clear that observation statements do not simply express the "given". They always transcend "pure perceptions" in that they make use of the conceptual pattern of the language employed. Observation statements are thus influenced by the conceptual structure of the language in which they are framed.

Imagine, for example, that a language L_1 contains a wider variety of color predicates than another language L_2. Take two persons using L_1 and L_2, respectively, and looking at two sheets of paper that are tinted in a slightly different fashion. It may well occur now that the observation reports of the two don't agree. The first one may ascribe distinct color predicates to each sheet, whereas the second one may describe both sheets by the same color predicate.

Logical Empiricism acknowledges such a language dependence (cf. Carnap 1936, 92–94). It is maintained, however, that observation statements do not rely on physical laws. They are independent of any theoretical interpretation. This implies that a given statement is by its very nature and without regard to the particular context in which it is used either observational or theoretical in character. Being an observation statement is not a two-place predicate, i.e., being observational with respect to a certain theory or with respect to a certain context; it is a one-place predicate instead. This is what the universality claim amounts to. Observations can be conceptualized without invoking laws. Consequently, the relation between theory and evidence, as conceived in the double-language model, is unidirectional. Theories are based on facts and not the other way around. This means, in sum, that the double-language model entails the denial of the theory-ladenness of observation.

2. THREE TYPES OF THEORY-LADENNESS

§ 1. The claim of the theory-ladenness of observations means that observations, observation statements or measuring procedures are influenced (or even determined) by theoretical assumptions or background knowledge. This claim may be entertained in three senses. It refers, first, to the influence of theories on the observations themselves, that is, on perceptual experience. I call this phenomenon theory-ladenness in the *perceptual* sense. A somewhat weaker claim is that there exists an influence of theories on the linguistic expression of observations. I refer to this peculiarity as theory-ladenness in the *contextual* sense. A third assertion is that the procedures applied in observation and measurement make use of theories themselves. I call this feature theory-ladenness in the *measuremental* sense. Let me now give a brief outline of these three types of theory-ladenness.

§ 2. Perceptual theory-ladenness was first systematically described by Norwood Hanson. In order to make his claim plausible, Hanson drew on some psychological theories, especially *gestalt* psychology. Hanson's point was that in perceiving we always apply some categories and ordering principles to the sensory input. A change in this cognitive apparatus thus entails a change in perception. It is clear, to be sure, that people with different cognitive settings may have the same retinal images. But in seeing more is involved than having retinal images; seeing is a matter of being aware of something. As it is demonstrated by *gestalt*-switch figures, two persons (or one person at different points in time) may be looking at the same configuration of lines and nonetheless see different things.

In such cases the elements of the two persons' perceptual experience are identical but these elements are ordered in a different fashion. What a person is aware of is determined by this organized set of elements. So, if different principles of perceptual organization are used, the same things are seen differently. Strongly held beliefs also belong in the class of perception-organizing principles. Imagine, as one of Hanson's examples runs, the geocentrically minded Tycho and the heliocentrically oriented Kepler sitting next to one another and watching the dawn. In spite of the assumed identity of their retinal images they don't see the same things happen. Tycho sees the sun rising whereas Kepler sees the horizon dipping (cf. Hanson 1958, 5–24).

In the same vein, Thomas Kuhn claimed that there is no neutral sensory experience and that theory intrudes into perception. Like Hanson, Kuhn invoked some experiments on and cognitive theories of perception so as to make plausible that a person's belief system may alter or falsify his or her perceptual experience. Applied to scientific observations this means that paradigm changes may induce perceptual changes. Scientists separated by a scientific revolution live in distinct perceptual worlds.[2]

§ 3. We may leave it undecided here whether this strong Hanson–Kuhn thesis of perceptual theory-ladenness is indeed true. For the purpose of the present study the two remaining types of theory-ladenness are more relevant. So let's turn to contextual theory-ladenness. Contextual theory-ladenness is an immediate consequence of the *theoretical context account* of meaning which has now largely supplanted the double-language model. According to that view, the meaning of a word is fixed by the rules which govern its use. In short, meaning is use. As regards scientific concepts, the relevant rules are the laws of the pertinent theory. A scientific concept gains meaning by its integration into some corpus of laws. For instance, the meaning of concepts referring to quantities like electric or magnetic fields is determined by the laws of electromagnetic theory. Understanding such concepts requires understanding the corresponding theory. It is the lawful relations which connect a concept with other concepts that specify this concept's meaning (cf. Hanson 1958, 61–62; Putnam 1962, 42–54; Papineau 1979, 35–45).

Contextual theory-ladenness emerges from the application of this general semantic approach to the meaning of observation reports. The claim is that the meaning of observation terms too is influenced by their theoretical context. This claim can be best illustrated by considering a change in the relevant laws. Suppose the term "magnetic field" is associated with the empirical indicator "fixed orientation of a magnetic needle". The discovery of electromagnetic induction then establishes a theoretical connection between electric and magnetic fields. Changing magnetic fields generate electric fields and the latter can be detected by the electric currents they produce, that is, by means of an ammeter. In addition, the Hall effect (i.e., Lorentz's force law) can be invoked to measure the intensity of magnetic fields. In this case the intensity is determined using a voltmeter. This example shows that changes in the theoretical context alter the conditions for applying the term "magnetic field" to experience. We can now decide about the presence of a magnetic field by means of an ammeter or a voltmeter and with no magnetic needle present. This term is now applied in situations in which it could not be applied previously. Hence the rules governing its use have changed and this amounts to a change in meaning.

§ 4. Systematically speaking, the context theory leads to the following interpretation of the observation language. First, every language induces some similarity or dissimilarity relations among the objects it describes. Suppose a language contains the concepts "elm tree" and "oak tree". In applying these concepts a number of actual trees are grouped together, say, as elm trees and separated from the class of oak trees. A conceptual pattern thus imposes a structure of equivalence classes on its domain of application. Second, several distinct indicators are in general suited to determine to which equivalence class a given object belongs. The assertion that a particular tree is an elm tree rather than an oak tree may be based on the examination of size or shape of the leaves or the structure of the bark or what have you. The presence of either one of these properties justifies the application of the corresponding predicate. This means that it is a cluster of properties – and not a single property – that is critical for the application of a concept. It follows from this, third, that these property clusters must possess a factual basis. Underlying the clustering of indicators is an empirical claim, namely, that the single items of this cluster always (or almost always) co-occur. The existence of an object with oak leaves and elm bark is excluded. A conceptual system thus generates a taxonomy as to which features are co-occurrent in the actual world and which are not. And this taxonomy constitutes an important element of the meaning of the concepts involved (cf. Kuhn 1983, 680–683; Hoyningen-Huene 1990, 485–487).

Let's now apply these considerations to the observation language. According to Logical Empiricism, the meaning of an observation term is fixed by linking an empirically accessible phenomenon to it. By virtue of this linkage the term is defined operationally, and all the additional statements in which it may enter are factual in nature. In this vein, Reichenbach distinguishes between the coordinating definitions which are analytic and the remaining statements which are synthetic (cf. Reichenbach 1928, 14–19; Glymour 1980, 19–20). But if meaning is determined by conceptual integration, this distinction collapses. On the context theory, all empirical indicators associated with a concept are on the same footing. Consequently, a whole cluster of empirical properties is involved in determining the meaning of a concept. And the important point is, then, that this property cluster is established by means of theoretical laws and is thus theory-dependent.

In the magnetic-field example it is electromagnetic theory that induces an equivalence relation between *prima facie* distinct empirical phenomena (such as magnetic-needle orientations and ammeter or voltmeter readings). They all are indicators of the same theoretical state. Temperature is another case in point. Temperature can be measured using a usual mercury thermometer or with the help of a resistance thermometer (which makes use of the fact that the specific resistance of metallic wires is temperature-dependent) or by analyzing the radiation emitted by a body etc.

Accordingly, "magnetic field" and "temperature" are theoretical terms in the Logical-Empiricist sense. Nonetheless, in usual scientific practice these terms are used in observation reports. "The temperature outside is 15° centigrade" would hardly qualify as a theoretical statement. The context theory stresses this appearance of theoretical terms in observation statements. The claim is, then, that in describing an observation in this way the pertinent theory is applied implicitly. It is applied by tacitly assuming that the observation statement could be based on a variety of procedures that are equivalent on theoretical grounds.

The structure of the observation language is thus dependent upon the way in which the relevant theory portrays its domain of application. This implies that conceptualization and interpretation cannot neatly be separated. In fact, Logical Empiricism had already envisaged this feature as a consequence of admitting multi-indicator concepts. But since multi-indicator concepts were supposed to occur only in the theoretical language, the impact of this insight did not extend to the observation language (cf. Nagel 1961, 82–83). On the context theory, however, observation concepts are multi-indicator concepts too, or put the other way around, observations are rendered in terms of theoretical concepts. Hence the dependence of description on theoretical explanation appears already on the observational level. The context theory thus entails that conceptual description and theoretical interpretation are strongly intertwined. And this is why a Logical-Empiricist observation language does not exist – at least it is not part of scientific practice.

§ 5. The third aspect involved in the theory-ladenness of observation is measuremental theory-ladenness. Measuremental theory-ladenness is supposed to express the fact that in measuring physical quantities one often has to rely on physical laws. Consider, for example, an astronomer who wants to determine the position of a radio star. He may stare at the star-speckled sky as long as he wishes without ever coming across the object of his interest. What he needs is the assistance of a suitably designed instrument, i.e., a radio telescope. The functioning of such an instrument is described by a theory, in this case, radio optics. Interpreting a signal peak of the registering device as a radio pulse originating from a cosmic source obviously presupposes, among other things, the availability of a theory about how the detecting device actually operates. The same holds with respect to a microbiologist who wishes to unveil the structure of a cell. When she uses an electron microscope to accomplish this task, she uses, in fact, the laws of motion of electrons in electromagnetic fields. To give a final example, consider a particle physicist who wants to test a theoretical model which predicts the mass values of certain elementary particles. In order to test such a prediction, the physicist may use a mass spectrometer. In such a device charged particles are exposed to the action of static electric and magnetic fields and by applying again the laws of electrodynamics the charge-to-mass ratio of particles can be derived from their observable path data.

These examples show that measuring in general involves a link between a state or quantity that is assumed and described by a theory, on the one hand, and the raw-readings of an instrument, on the other. The important point is that this link is established by and crucially dependent on some theory. Only by having recourse to such a theory, does a device become a measuring instrument; that is, only by virtue of such a theory, it is justified to claim that a certain state of that instrument corresponds to a certain state outside of it.

A note as to history. This role of theories in observation and measurement procedures was recognized and elaborated chiefly by Pierre Duhem and Imre Lakatos (cf. Duhem 1906, 153–156; Lakatos 1970, 22–24, 42–45). In addition, it was sketched by Reichenbach. Reichenbach's "indirect verification" virtually coincides with measuremental theory-ladenness (cf. Reichenbach 1938, 46–48; 53; cf. also Reichenbach 1932, 115). A note as to terminology. Theories that connect a theoretical state or quantity with an empirical indicator are called *observation theories* in the sequel.

This means, in sum, that observation theories are typically those theories which describe the operation of the instrument employed. Such theories allow one to track the causal chain leading from the physical event to be registered to the reading of the instrument. Without this theoretical reconstruction of the path the signal takes within the instrument, the interpretation of the output of the instrument as the representation of some external state is without justification. Measurements depend on the availability of such observation theories and are theory-laden in this sense. In the next section I take a closer look at this important peculiarity.

3. MEASUREMENTAL THEORY-LADENNESS: SOME DETAILS AND APPLICATIONS

§ 1. In this section I wish to expand on some features of measuremental theory-ladenness. The best place to begin seems to be the elaboration of an example. Let's envisage the detection of radioactive radiation by way of a solid-state counter. The question to be addressed is: On what grounds does such a device qualify as a reliable means for recording the presence, intensity and energy spectrum of radioactive radiation? The answer to this question relies on the physical characteristics of the radiation as well as on those of the device employed. The relevant property of the radiation is its ionizing effect. That is, it releases electrons from the molecules of a suitable material. An incoming particle generates a large number of oppositely charged particles (positive ions and negative electrons) in the detector material; and by applying an electric field these particles can be separated and recorded.

This rough scheme is implemented by means of a semiconductor diode. Such a diode consists of a junction between an N-type and a P-type semiconductor. N-type semiconductors contain impurities that easily give off electrons into the so-called conduction band; electrons moving in this band constitute an electric current. P-type semiconductors contain impurities that remove electrons from the so-called valence band. These impurities thus produce positively charged holes in this band, and the motion of these holes also represents an electric current. If these two materials are brought into contact, oppositely charged particles recombine in the contact zone and produce a layer of vanishing net charge. This depletion layer constitutes the radiation-sensitive area, and it can be expanded by applying a suitably directed electric field.

An ionizing particle entering the depletion layer generates a vast number of positive and negative charge carriers. By the action of the electric field these charged particles are separated and driven to their respective electrodes where they give rise to a current pulse. The electric field inside the detector thus serves to collect these particles. The number of charge carriers produced is proportional to the energy given off by the incoming particle in the detector material. Consequently, the pulse height is a measure of the total energy deposited by that particle in the counter, and the counter thus works as an energy spectrometer.

What we have here, in sum, is a link between the pulse height of a current in the instrument and the energy of an incoming ionizing particle. This link is established by physical laws. It is the band theory of solids (along with some basics from electrodynamics) on which the inference from pulse height to energy rests. Interpreting a pattern of current pulses as energy spectrum of the radiation prevailing outside is only justified by virtue of these laws and theories. This interpretation relies on using these theories as observation theories.

It appears from this discussion that observation theories do not simply connect a theoretical state to some empirical indication. Rather, they connect *specific values* of the two quantities with one another. The observation theories under consideration license the inference from a particular pulse height to a particular energy value. This job is more demanding than the mere tying together of two states. The latter task can also be fulfilled by employing a theory as a *criterion of adequacy* for appropriate measuring procedures. That is, these procedures are selected on the criterion that their results fit the theoretical expectations – at least in clear-cut cases. Consider a would-be detector that remains mute when placed next to some (chemically identified) radioactive material. In that event we would start to harbor doubts regarding the trustworthiness of the apparatus. In sum, using a theory as criterion of adequacy amounts to placing demands on every process that is to qualify as a legitimate indicator process of the theoretical quantity in question. Using it as an observation theory, by contrast, means deriving a relation between particular outcomes of the indicator process and particular magnitudes of that quantity.

Another important feature of observation theories can be gathered from this example. Reliable measurements may be vitiated by the presence of *perturbations*. Distorting influences may interfere with the reliability of the detection process. For instance, if some particles simply pass through the detector without triggering any reactions, the intensity of the radiation is misrecorded. This mistake can be avoided by resorting to additional laws. Nuclear theory gives an estimate of the reaction probability and thus of detector efficiency. The actual intensity can be inferred, accordingly, from the measured intensity and the detector efficiency.

A second source of error is that a particle fails to deposit its whole energy in the counter and escapes after having brought about some ionizations. This effect leads to a distorted shape of the energy spectrum. The same result occurs if some of the charge carriers produced are trapped by impurities on their way to the electrodes. In that case not all the charged particles generated are actually collected. Both kinds of perturbations lead to incorrectly low energy values. These disturbing influences can be avoided, first, by making the depletion layer thick enough so that even high energy particles are stopped in it, and, second, by making sure that no charge traps are present in the detector material.

These considerations make it clear that observation theories usually hint at or circumscribe possible causes of mistaken measurements. Since they account for the measuring process, they often allow one to appraise where and how the measurement might go wrong. There are two general strategies at hand for repairing such flaws. These strategies are, first, *correction of the perturbations occurring* and, second, *realization of the undistorted case*. The two procedures described above exemplify these strategies.

The non-ideal detector efficiency leads to incorrect data about radiation intensity. This problem is overcome by quantifying the distorting effect (i.e., by evaluating the detector efficiency), and by using this estimate as a basis for calculating the undistorted value. That is, the magnitude of the perturbation is determined, and this result is used for correcting the obtained raw-value. The alternative to eliminating the impact of a distortion is eliminating the distortion itself. This is done by realizing the ideal or undistorted case, i.e., by removing possible sources of error (i.e., insufficient extension of the sensitive layer and the presence of electron traps). If this can actually be achieved the results of the measurements don't have to be corrected but are reliable as they stand. It is clear that both these general strategies make essential use of observation theories.

This means, in sum, observation theories connect a theoretical state with an empirical phenomenon by exposing and analyzing the chain of processes leading from the first to the second. Such theories thus provide an explanation why that phenomenon constitutes a reliable indicator of the state in question. Moreover, observation theories contribute to specifying the limits of reliability by bringing to light possible perturbing influences that may vitiate the trustworthiness of the measurement. And they help to reestablish reliability using a two-track strategy. First, rectification of the distortion through the theoretical calculation of its magnitude and impact. Second, specification of means suited to avoid its occurrence. In short, observation theories explain the reliability of certain measurement procedures and account for their possible perturbations.

§ 2. I would like to illustrate the role of observation theories in correction procedures by two examples from Isaac Newton's theory of gravitation. It was one of the touchstones of Newtonian celestial mechanics that it succeeded in accommodating the observed planetary positions. But in this context the problem arose that as a result of the refraction of light by the atmosphere, the positions as directly observed do not coincide with the actual ones. Light rays are slightly deviated from their straight paths when they enter and pass through the atmosphere, and the planet thus appears displaced. What one needs here is an optical theory that is suited to quantify the magnitude of this distorting effect. By using optics as an observation theory, the raw data can be corrected so that the true planetary positions may be inferred. In Newton's time this correction problem was indeed urgent, and Newton had a discussion with the astronomer John Flamsteed about how to perform this correction adequately.[3]

The same trait can be recognized by considering the measurement of the gravitational acceleration of the earth with the help of a pendulum. This quantity can be derived in principle from the periods of oscillation of a pendulum. However, such a measurement is distorted by the influence of the centrifugal force which is an effect of the earth's rotation. This centrifugal force gives rise to an apparent reduction of the gravitational acceleration that varies with geographical latitude. Newton came across this problem in his attempt to determine the actual shape of the earth. If the daily rotation brings about a noticeable flattening of the earth's body, the gravitational acceleration should be greater at the poles than near the equator. For a body located at a pole is closer to the earth's center and hence is attracted more strongly. This effect is veiled, however, by the action of the centrifugal force on the measuring device, i.e., on the pendulum oscillations. This distortion vitiates a reliable determination of the true gravitational acceleration. Newton tackled this problem by assuming a spherical shape of the earth and by calculating the magnitude of the distorting effect on that basis and with the help of mechanical theory. This procedure results in corrections of the raw data obtained from pendulum measurements, and these corrected data can then be used for determining the actual shape of the earth (cf. Newton 1726, 424–430). The conclusion Newton was after could only be based on an examination of the true weights of bodies, and these true weights are only accessible by resorting to a correction procedure.

It was probably Duhem who first emphasized systematically the need for observation theories as a safeguard against distortions. Observation theories are required in order to correct distorted evidence. True, there are some cases, Duhem argued, in which the existence of a link between the outcome of an observation or measurement and the state it is supposed to represent can be verified by direct inspection. Even in these cases, however, the correction of perturbations demands application of a theory. Duhem's example is the magnifying glass. It can be confirmed by unaided observation that a magnifying glass is an at least approximately reliable instrument for producing enlarged images of objects. So its use can be justified without invoking any theory.

But even in this case, recourse to an observation theory is inevitable if distortions are taken into account. Chromatic aberration leads to the appearance of colored edges of the objects seen; these objects are surrounded by fringes in the spectral colors. Only a theory of dispersion puts us in a position to construe this effect as a distortion, i.e., as an artifact produced by the instrument. After all, it might have been the case otherwise that these colored fringes actually belonged to the objects themselves. In reality, all objects are surrounded by tiny halos, and with the help of the magnifying glass we have managed to bring to light this remarkable property. Such an interpretation can only be ruled out by resorting to optical theory (cf. Duhem 1906, 153–154).

The point of this example is that even in cases in which a direct comparison between the object and the data recorded is possible, we are still in need of observation theories. By specifying distortions these theories imply what the pure state is like. Observation theories distinguish between what is part of the sought-for effect and what is not. This feature is even more marked in those cases, usually encountered in science, in which the only access to an object leads through theory-based measurements. In such cases the whole justification of an experimental set-up turns upon observation theories.

§ 3. Let me illustrate this latter dependence by one more example. This example is supposed to show that the interpretation of a particular set of data as indications of a specific theoretical state is indeed critically dependent upon the observation theory used. Whereas the examples discussed in § 2 focused on the importance of observation theories for treating distortions, the present example is intended to underline the significance of observation theories for establishing a relation between theory and evidence in the first place. I sketch a real-life case from the history of chemistry in which distinct observation theories were applied to the very same laboratory data with the result that one and the same empirical record was construed as representing completely different theoretical states.

The relevant empirical data are the relations between the combining weights of substances reacting chemically with one another. Let's consider the interpretation of these data as given within the framework of affinity theory, on the one hand, and John Dalton's atomic theory on the other. Affinity theory was the generally accepted 18th-century account of chemical reaction and chemical bond.[4]

Its fundamentals were developed by Newton. Affinities were conceived as short-range forces acting between chemically active particles, and their strength was assumed to vary from substance to substance. Moreover, affinity forces can be saturated, that is, their strength is diminished by attracting further particles. This means, that one particle of a substance A can attract only a limited number of particles of a substance B. If the limit of B-particles is reached, the attractive power of A is exhausted. Beyond this so-called point of saturation the affinity forces are no longer effective.

On the basis of this theory, affinities can be measured as follows. The stronger the affinity of an A-particle to the B-particles is, the more B-particles are bound by one A-particle. Furthermore, during the whole 18th century it belonged to the unarticulated parts of conventional scientific wisdom that all fundamental particles (i.e., "atoms") had the same weight (cf. Carrier 1986b, 10–11). On the basis of this tenet, different capacities for attracting particles can be directly translated into different combining weights. That is, the stronger the affinity of A to B, the greater is the weight of B necessary to saturate a given weight of A, and this saturating weight of B is evidently indicated by the combining weight of B. Accordingly, gravimetric combining proportions count as quantitative indicators of affinity strengths.[5]

Within the framework of Dalton's atomic theory, published in the first decade of the 19th century, things are completely different. Dalton assumed that a small number of atoms combined to form a "compound atom" (i.e., molecule). Moreover, he believed – in contrast to the 18th century tradition – that atoms of different kinds possess different weights. Weight is a property that is characteristic of the atoms of an element. Within this approach a novel research problem was created, namely, the problem to determine empirically the atomic weights of various elements.

At this juncture combining weights enter the stage. Dalton thought that the combining weights were suitable as indications of the atomic weights. Take three elements A, B, C and suppose that they combine to molecules AB and AC, respectively. Assume furthermore that the weight of B reacting with a fixed weight of A is just double the corresponding weight of C. In that case it seems plain that the atomic weight of B is twice the atomic weight of C. There is, however, a difficulty involved here. The inference from the combining proportions to the atomic weights only works if the molecular formulas are known. Assume, for example, that the correct molecular formula is AC_2 rather than AC. In that case the atomic weight of B is four times the atomic weight of C. And what makes this problem really hard to solve is that in Dalton's time there was no independent way of ascertaining empirically the molecular formulas involved.

So, we have a situation in which only one type of empirical data (namely, combining weights) was available to determine two distinct theoretical states (namely, atomic weights and molecular formulas). These theoretical states are reciprocally dependent upon one another; that is, if one of them is altered the other one is to be adjusted accordingly. This leads to the result that there exist a multitude of conjunctions of such theoretical states each of which is equally suited to fit a given set of combining-weight ratios. The theoretical states are thus underdetermined by the available evidence.

Dalton tried to overcome this problem by adding one more postulate to his theory, namely, the "rule of greatest simplicity". If only one compound of two elements A and B is known, it is posited that the molecular formula is AB. If two or three compounds exist, it is supposed that the molecular formulas are AB, A_2B (or AB_2) or AB, A_2B, AB_2, respectively. By resorting to this rule, the molecular formulas can in most cases be determined unambiguously, and this allows for the derivation of atomic weights from the combining proportions.[6]

On the basis of the two distinct theoretical approaches the very same empirical relations, namely, the ratios of combining weights, are thus interpreted as indications of completely different theoretical states. If a substance A reacts with two substances B and C, and if the weight of B combining with a fixed weight of A is greater than the corresponding combining weight of C, then this means on the basis of affinity theory that the attraction from A-atoms to B-atoms is stronger than the attraction from A-atoms to C-atoms. By contrast, on the basis of atomic theory the same findings allow for the conclusion that the atomic weight of B is greater than the atomic weight of C. This shows that applying different observation theories to the same set of data gives rise to diverging interpretations as to what these data actually represent. It is such observation theories that establish the link between theory and evidence.

§ 4. These considerations show that it is the theoretically interpreted and enhanced evidence that is scientifically significant. In sum, observation theories serve the double purpose of, first, singling out procedures that are suited to register some theoretically specified states or processes, and, second, preventing distorting influences from thwarting the measurement. In other words, they serve the double purpose of, first, making measurements *possible*, and, second, making them *reliable*.

The role of observation theories as an intermediate link between theoretical states and measurement outcomes gives rise to an important peculiarity, namely, the occurrence of a particular type of systematic observational errors. Such errors are not the result of simple slips but arise from the flawless application of a flawed observation theory. It is precisely this feature that appears in the combining-weights example. On the basis of affinity theory, the inference from combining weights to affinity strengths is perfectly sensible. Nonetheless, it amounts to a systematically wrong interpretation of the data. Systematic observational errors of this type cannot be removed by carrying out the observations more carefully and more accurately. It can only be removed by improved theorizing.[7]

There is a further aspect involved here which deserves notice. No natural or principled difference exists between explanatory theories and observation theories. That is, in different experimental contexts the same theory can play the role of an observation theory, for one, and of an explanatory theory, for another. Take particle physics as an example. Assume that a particle theory makes predictions about the half-life, spin etc. of some kind of particles. These predictions can be tested by relying on those theories that describe the functioning of the relevant detector devices. Hence, our particle theory figures as explanatory theory in this connection.

Let's now consider an astrophysical model of the sun. The flux intensity of solar neutrinos constitutes an important test indicator for every such model. This flux intensity, however, can only be measured in an indirect fashion. Namely, a neutrino can only be detected by way of certain capture reactions. That is, the neutrino is captured by another particle which decays subsequently, and the decay products are experimentally detectable. The point is that suitable capture reactions and the nature of the decay products are specified by particle physics. It is particle theory on which the interpretation of the occurrence of a certain decay product as indication of a captured neutrino relies. In this context, the particle theory thus plays the role of an observation theory with respect to the explanatory astrophysical model.

This implies that there is no universal, i.e., context-independent, demarcation between observation language and theoretical language (as it was assumed within the double-language model (cf. I.1.§3). The same statement, e.g., the assertion that a certain particle reaction has taken place, may count as theoretical explanation in one context, and as empirical observation in another. The absence of a universal demarcation between the two linguistic levels is due to the fact that observation statements indeed rely on physical laws. Measuremental theory-ladenness is thus in conflict with the Logical-Empiricist universality claim. On the other hand, the distinction between the two levels still makes sense with respect to particular test situations. In every particular empirical test, the evidence to be explained can be distinguished from the explanatory account.

CHAPTER II

THE COMPLETENESS OF THEORIES

1. EINSTEIN–FEIGL COMPLETENESS INTRODUCED

§ 1. In all the examples hitherto discussed explanatory theory and observation theory were separate theories. A theory T_1 served as observation theory for a theory T_2 distinct from T_1. For example, the band theory of solids was employed as observation theory for particle theory (cf. I.3.§1); optics and mechanics were used as observation theories for the explanatory theory of gravitation (cf. I.3.§2). However, there also exist instances of a different kind. It may happen that a theory figures as its *own* observation theory. In such a case a theoretical term introduced and specified within a theory is connected to an empirical procedure by virtue of the laws of the very same theory. Let's first turn to an example in order to get a clearer idea of how this is to be understood.

Consider the measurement of current intensity by means of a moving-coil galvanometer. In such a device, the current whose intensity is to be determined passes through a coil; this coil is suspended from a spring and is located within a magnetic field generated by a suitably shaped permanent magnet. The current in the coil produces a magnetic dipole moment which interacts with the magnetic field. This interaction gives rise to a torque that twists the spring. The coil is thus rotated and eventually assumes an equilibrium position in which the magnetic torque is balanced by the elastic torque of the spring. The resulting angle of rotation expresses the current intensity present.

Current intensity is a theoretical quantity that is introduced within the framework of electromagnetic theory. And it appears from the just-given theory-laden description of the galvanometer that the relation between this electromagnetic quantity and its observational indicator "angle of rotation" is established (in large part) by electromagnetic theory itself. After all, the quantities employed in that description (such as magnetic field and magnetic dipole moment) are likewise specified within that same theory. That is, the nature and interaction of these quantities is determined by laws that are themselves part of electromagnetic theory. The functioning of a moving-coil galvanometer can be accounted for on the basis of these electromagnetic laws. Their application serves to analyze the operation of a device that is supposed to measure a quantity appearing in these same laws. This means that electromagnetic theory explains why the instrument discussed is associated with one of its own theoretical concepts. By giving a theoretical analysis of this procedure in its own terms, the theory furnishes a justification for using it as a measuring device.

In addition, electromagnetic theory also allows for a theoretical treatment of some systematic distortions. Consider a further aspect of the galvanometer example. If a galvanometer is actually to be used for measuring current intensity, it has to be inserted into the corresponding electric circuit. But by doing this one inevitably brings about a decrease in the current intensity to be evaluated. This decrease is due to the internal resistance of the galvanometer which enlarges the overall resistance of the circuit and consequently contributes to a reduction of the current intensity. Applying the galvanometer thus generates an alteration of the sought-for quantity. Electromagnetic theory specifies this distorting effect; after all, I have just described this perturbation within its conceptual framework. Moreover, it opens an avenue for its correction. It is possible, namely, to determine separately the internal resistance of the galvanometer and to calculate on that basis the undistorted value of current intensity (the relevant details are explained in II.5.§1).

This presentation entails that the two characteristic features of observation theories are in fact present here. Electromagnetic theory establishes and justifies the link between a theoretical quantity and its empirical indicator, and it provides the basis for a correction procedure of systematic errors. This means that electromagnetic theory indeed operates as a full-grown observation theory, and the noteworthy aspect is that it does so with respect to one of its own theoretical concepts. The conclusion is that electromagnetic theory really figures here as its own observation theory.

§ 2. The notion that a theory can establish on its own the relations between the states assumed within its framework and their empirical indicators was first circumscribed by Einstein in 1949. He gave it the form of a methodological postulate. Einstein objected that Special Relativity must simply require the characteristics of ideal rods and clocks which the theory deals with, and he regarded this feature as a methodological shortcoming. As he observed, Special Relativity introduces two kinds of physical things, i.e., (1) measuring rods and clocks, (2) all other things, e.g., the electro-magnetic field, the material point etc. This in a certain sense is inconsistent; strictly speaking measuring rods and clocks would have to be represented as solutions of the basic equations (objects consisting of moving atomic configurations), not, as it were, as theoretically self-sufficient entities (Einstein 1949, 59; cf. also Einstein 1920, 662).

That is, the theory of the atomic constitution of matter ought to contain or imply a theory of the measuring instruments. From a comprehensive theory of matter those properties of the behavior of rods and clocks should be derivable which can only be postulated within Special Relativity. The latter theory is thus construed as an account of the measuring instruments and it is required that it be a consequence of a more basic explanatory theory rather than be formulated as an independent theoretical approach.

In a similar vein, Feigl argued in 1950 for the methodological superiority of theories containing theoretical terms by claiming that only a theory of this kind could possibly manage to treat its observation basis – perturbations included – with conceptual means belonging to this theory itself. If self-sufficiency of this sort has been accomplished, a veritable "Copernican turn" has occurred. This turn

consists in the derivation, with corrections coming from the theoretical scheme, of the peculiarities of the very basis of confirmation [...] As long as a science has not attained a very high level of explanation the process of indication may not be deducible from theoretical premises [...] [T]his corresponds to the theory of acids and bases in chemistry at a time when the process involved in the well-known litmus paper test was not itself logically derivable [from chemical theory ...] The behavior of such thermometric substances as alcohol or mercury was not theoretically deducible until the kinetic (molecular) theory of heat put these indicator processes on a par with countless other thermodynamic processes *as interpreted* on the micro-level [...]The Copernican turn then consists in relating the observer to the observed, the indicator to the indicated, – not epistemically, – but so to speak cosmologically. What epistemically must be looked at as the confirmation bases of the hypothetical construction, will in the full-fledged theory be given a place within the cosmos of which the theory treats (Feigl 1950, 40–41).

Accordingly, the measuring procedures that are used within a theory for implementing its concepts into experience should be part of that theory's domain of application. This requirement is explicitly extended to the theoretical treatment of distortions; as Feigl put it, the latter's corrections should "come from the theoretical scheme".

The gist of these two quotes can be cast in the claim that it is, first, possible and, second, desirable, that a theory provides a basis for the justification, theoretical analysis and theoretical correction of the indication processes on which it relies. Evidently, the substance of this claim roughly coincides with the peculiarity outlined in § 1. Hence I refer to this peculiarity as *Einstein–Feigl Completeness*. That concept says, consequently, that the respective theory operates as its own observation theory.[1]

The methodological virtue of this capacity is obvious. If a theory is conceptually autonomous in this sense, it achieves on its own what its incomplete counterpart accomplishes only through external assistance. Complete theories thus get high rankings on the parsimony scale, and this distinction is one of the most widespread criteria for methodological quality (cf. V.4.§3).

As already mentioned in I.3.§1, it is characteristic of observation theories that they are actually used for *inferring* or calculating the *specific values* that a theoretical quantity assumes in particular cases. It is natural to transfer this feature to the case of complete theories. This means, it is *not* sufficient for the completeness of a theory that it merely supplies *conditions of adequacy* for possible indicator processes. Giving mere conditions of adequacy only amounts to postulating that processes with the requisite properties exist; it amounts to selecting prospective instruments according to the agreement of their outcome with the theoretical anticipation.

As an example for that latter use of a theory consider Hendrik Lorentz's response to the Michelson–Morley experiment. On the basis of Lorentz's electron theory, that experiment should have given a positive result indicating the earth's motion with respect to the ether. In fact, however, Michelson Morley came up with their famous null result. In the wake of that surprising event, Lorentz conceived his "contraction hypothesis" that aimed at removing the experiment from the class of possible indicator processes for the supposed ether drift. This was accomplished by the introduction of a distorting factor (the length-contraction of moving bodies) whose effect was precisely to compensate the influence of the sought-for process. The experiment thus did not indicate in reality what it was previously supposed to indicate; it only indicated the combined effect of ether drift and Lorentz contraction. In this case the explanatory theory is invoked as criterion of adequacy for the relevant measuring procedure. Faced with an anomaly that theory is retained and the observation theory changed so that agreement with experience is reestablished.

It is clear that any theory whatever can be used as criterion of adequacy. Such a use is not based on a special feature of a theory but merely testifies to a strong commitment to that theory. It only shows that confidence in the theory outweighs confidence in the appropriateness of the measurement procedure. So, if completeness is to be a philosophically interesting concept, complete theories should do more than restrict the range of admissible observation theories. Hence, a theory offering mere criteria of adequacy for suitable indicator processes does not count as complete. What is required is theoretical analysis of such processes and not merely their selection.

§ 3. Einstein–Feigl completeness means, accordingly, that the explanatory theory itself is contained within the class of the observation theories relevant to it. It is not ruled out by this characterization that there are additional relevant observation theories. It may be the case, for instance, that one theoretical state is associated with several empirical indicators and that the explanatory theory itself figures only in one of these links. This is precisely what happens in the galvanometer case. Whereas the electromagnetic method for measuring current intensity is based to a great extent on electromagnetic theory itself (as explained in § 1), there exist additional measuring procedures that are not. The operation of electrochemical or electrothermic methods for measuring current intensity cannot (or can only to a small degree) be explained by recourse to electromagnetic theory. The operation of these methods lies outside that theory's scope.

These considerations suggest that there exist degrees of completeness. A theory may establish the empirical links of its theoretical terms to a greater or lesser extent. Let's restrict our attention, for a start, to one particular theoretical state and its empirical implementations. In that case, the following four possibilities may arise.

(1) The explanatory theory is *necessary and sufficient* for the analysis of the pertinent measuring procedures. In that case, every empirical implementation is comprehensively accounted for by the corresponding theory.
(2) The explanatory theory is *necessary but not sufficient*. This means, every measuring procedure relies in some way on at least one law of the corresponding theory. In addition to these laws, however,

other laws or theories are required for a comprehensive analysis of the relevant devices.

(3) The explanatory theory is *sufficient but not necessary*. In that case, at least one of the implementing procedures can be comprehensively accommodated within the theory in question, and another one can be analyzed without any recourse to that theory. That is, there is one track from theory to evidence for which that theory is necessary and sufficient and another track for which it is neither necessary nor sufficient.

(4) The explanatory theory is *neither necessary* for the analysis of every procedure *nor sufficient* for the analysis of any particular one; but it is necessarily involved in establishing at least one of the relevant links to experience.

Let me put terminological labels on these distinctions. I speak of *strong* Einstein–Feigl completeness if the explanatory theory is *necessary* for the analysis of *every* measuring procedure tied to the theoretical state under consideration. This encompasses the first and the second of the options just described. If, by contrast, at least two independent alternative treatments of the relevant observational procedures can be given, I speak of *weak* Einstein–Feigl completeness. This comprises the third and the fourth of the options outlined. The complementary concept to completeness is *simple theory-ladenness*. Simple theory-ladenness is supposed to mean that a theory is neither necessary nor sufficient for the analysis of any of the relevant instruments. That is, all pertinent observation theories are distinct from the explanatory theory, and the scopes of the former do not lie entirely within the domain of application of the latter (so that the observation theories can be tested independently of the explanatory theory). Measuremental theory-ladenness is thus split up into the two mutually exclusive sub-concepts "simple theory-ladenness" and "completeness".

The just-given classificatory scheme can be transferred analogously to the ordering of self-performed corrections. That is, a theory may be sufficient or necessary, in the combinations specified, for performing the corrections requisite for a reliable indication of a state introduced in that same theory. And it may be sufficient or necessary with respect to at least one such correction or with respect to all of them. There is thus a considerable range of possibilities opened up here. But fortunately enough, most of them need not concern us.

§ 4. Let's examine briefly which of these categories of completeness the electromagnetic case referred to repeatedly actually belongs to. I focus on the aspect of establishing evidential ties and leave aside for the moment the further aspect of supplying corrections. To state it right at the start, electromagnetic theory instantiates the fourth option with respect to current intensity. The theory is indeed necessarily involved in the account of the galvanometer but it is not sufficient for that purpose. For in addition there is a mechanical law entering that procedure, namely, the law describing the twisting of the spring as a response to the magnetic torque (see § 1). This rotational analog to Hooke's law is requisite for a comprehensive account of the measuring process. Electromagnetic theory thus necessarily figures in one of the links tying current intensity to experience, but does not establish that link entirely on its own.

On the other hand, electromagnetic theory is not necessary *simpliciter* for establishing these links. Besides the galvanometer method there exist procedures that do not rely on electromagnetic theory as observation theory. As already mentioned, it is possible, for instance, to calculate the current intensity from the mass deposited per unit time in an electrolytic decomposition process. These calculations are based on lawful relations concerning the atomic constitution of matter and make use of quantities such as molecular weight and Avogadro's constant. Electromagnetic theory at most enters as a criterion for the adequacy of that procedure. Accordingly, there is at least one measuring procedure in which electromagnetic theory does not figure as observation theory.

Finally, there is no pertinent measuring apparatus – at least there is no one known to me – that can be comprehensively analyzed by electromagnetic theory alone. For the purpose at hand electromagnetic theory is thus not sufficient either. And this means, in sum, that electromagnetic theory is complete with respect to current intensity in the weak sense (4) only.

There exists one more aspect or scale on which the degree of completeness may vary. Till this juncture, I have picked but one theoretical concept and examined the possible relations of that concept to experience. But a theory clearly contains more than one concept. A theory may thus be complete or incomplete with respect to either one of the concepts it introduces or defines. The *qualitative* differences in the concept of completeness (as outlined above) are thus supplemented by *quantitative* differences. A theory may be complete with respect to one of its concepts, or all of them, or some intermediate number.

There is some more fine-structure to be added. Up to now, I used the term "theory" in a broad and loose sense. Actually, a theory is in general a deductively organized set of laws or hypotheses. This suggests that the concept of completeness should be ramified further by taking into account which individual laws and particular hypotheses are actually employed. Let's reconsider the galvanometer case from this perspective. Assume we wish to determine the current intensity as it appears in Ohm's law. By making use of the galvanometer, we essentially apply the following three laws: a law linking the current to be measured to the magnetic dipole moment it generates, a law expressing the interaction between this quantity and the static, external magnetic field, and a law tying the magnetic torque to the mechanical twist of the spring. Ohm's law appears nowhere in this scheme. That is, in order to measure the quantity under consideration in one particular law, we need not have recourse to that law itself; other laws (mostly from the same theory) will do instead.

This can be expressed by distinguishing between *completeness on the theory level* and *completeness on the hypothesis level.* In the former case several laws from the same theory are applied for analyzing (and correcting) the pertinent measuring procedures. In the latter case, however, it is actually one law that is necessary (or even sufficient) for analyzing (or even correcting) the procedures suited to evaluate one of the quantities it itself contains. It is clear from the outset and can moreover be gathered from the example just given, that completeness on the theory level does not entail completeness on the hypothesis level. Rather, completeness on the theory level can dissolve into simple theory-ladenness on the level of the hypotheses actually employed.

§ 5. The situation becomes more complicated if we take corrections into account. As described in § 1, the application of the galvanometer leads to a decrease in the current intensity to be measured. In order to correct this distortion we have to determine the internal resistance of the instrument. This determination can be based on Ohm's law; we apply a given voltage to the instrument and measure the current intensity that emerges. The obvious difficulty is that this second measurement of current intensity is distorted in the same way as the original one that we wish to rectify. Accordingly, a vicious circle seems to surface here.

What we are actually dealing with at present are test procedures for particular hypotheses. Such hypotheses are tested by calculating the values the quantities occurring in them assume in particular cases.[2]

The result of the above discussion is that the test of Ohm's law is unproblematic if distortions are left out of consideration. But if they are included in the scheme a circle emerges that stems from the fact that the hypothesis to be tested is itself applied in the correction procedure. But this self-reference is precisely characteristic of potential completeness on the hypothesis level. So, if there is completeness of this sort, i.e., if a hypothesis guides the evaluation of one of the quantities occurring in it, a test of this hypothesis may be ruled out. If distortions are taken into consideration the quantities in question can apparently no longer be evaluated unambiguously.

Up to the present juncture, completeness seemed to represent a rather unproblematic conception. But now we must realize that a serious danger is lurking behind its attractive appearance. Theoretical self-sufficiency may issue right into a circularity which in turn may restrict considerably or even preclude the testability of the respective theory. In fact, it is these seemingly stark prospects for completeness that constitutes a major issue in the following considerations. If completeness actually were the innocuous beauty as which it was introduced, the project of studying completeness would already be finished. It's otherwise though. Hence, if a complete theory is to represent an asset rather than a burden, it has to be shown that these possible test problems can appropriately be handled.

Exploring the possible dark spots behind the elegant outfit calls for an additional terminological distinction. Completeness as introduced in this section is by definition a successful endeavor. A complete theory is defined so that the self-reliant transition from the theory's core premises to the relevant pointer readings can indeed be accomplished; and it is defined so that the necessary corrections can actually be effected in a theoretically self-sufficient fashion. Clearly, if we want to investigate the prospects and risks involved in that notion, a more neutral and less pre-empted terminology is requisite.

Accordingly, if there are indications that a theory operates as its own observation theory while leaving it open whether or not the possibly occurring circles can be overcome, I speak of *potential completeness*. And if it turns out at last that the self-referential traits characteristic of completeness lead us into a messy circle rather than to definite pointer readings, I speak of *failed completeness*. In order to avoid terminological clumsiness, however, I take the liberty of dropping the qualification "potential" in those instances in which no misunderstanding can arise. So I will sometimes speak of "test problems of complete theories" although this expression, strictly speaking, involves a contradiction in terms.

What I want to examine in the following is the question, whether completeness has in fact to be purchased at the price of serious test restrictions. Are the test circles that apparently go along with the endeavor to complete scientific theories really vicious in kind? What are the strategies, if there are any, to break up or otherwise block such circles? I tackle this problem by way of a detour. I begin by discussing a further ancestor (somewhat remote, to be sure) of the concept of Einstein–Feigl completeness, as developed here, namely, Joseph Sneed's concept of theoretical terms.

2. SNEED'S "PROBLEM OF THEORETICAL TERMS" AND THE TESTABILITY OF STRONGLY COMPLETE HYPOTHESES

In 1971, Sneed proposed a logical reconstruction of the characteristics of theoretical terms which bears a conspicuous resemblance to my conception of strong Einstein–Feigl completeness. Interestingly enough, Sneed claimed that the appearance of theoretical terms in his sense indeed restricts the testability of the pertinent theory. Examining the appropriateness of this claim leads to an improved understanding of the test problems that possibly arise in complete theories.

§ 1. I begin by briefly outlining the essentials of Sneed's "structuralist approach". In doing this, I confine myself to those of its aspects that bear somehow on the "problem of theoretical terms," i.e., on the difficulties of theory testing that Sneed believes he has identified. In fact, not many details of structuralism are requisite here since Sneed's concept of theoretical terms is largely independent of the remaining structuralist particulars.

The central idea behind the structuralist position exhibits strong ties to the so-called "semantic view" of theories: A theory is not to be conceived as a deductively ordered system of statements; rather, a theory should be identified by way of the objects and phenomena to which it applies. A theory is characterized by its domain of application. Of course, this should not – and in most cases cannot – be done by simply enumerating the members of that domain. It is to be achieved in a more general fashion instead, and that means, by specifying some relevant properties of the range of phenomena under consideration. These properties are then conjoined to an abstract predicate, and the empirical claim associated with a theory is that this predicate rightly applies to the domain of interest. The claim is that every particular case from among this domain instantiates the general properties specified by the predicate. The physical systems that indeed exhibit the required features are called the "models" of the respective theory.

The basic unit of the structuralist analysis of theories is the "theory-element". Theory-elements are the smallest structuralist entities to which an empirical claim can be associated. Theories in the intuitive sense are then reconstructed as networks of such theory-elements. These elements thus represent the basic form of a theoretical predicate. A theory-element is characterized, first, by its core which consists in turn of two distinct parts. One part expresses the conceptual framework of the theory-element, and this framework is characterized by the formal properties of the concepts used. The other part contains a number of laws (or one single law) formulated in these concepts. Whereas the first part is supposed to represent the conceptualization of a domain of intended applications, the second renders the factual claims associated with the objects in this domain.

To make things more perspicuous let me give an example of the core of a theory-element, namely, that of "classical collision mechanics". This theory-element deals with the collisions of two point-mass particles, and the relevant law is the law of the conservation of linear momentum. In that case the conceptual framework is specified as follows. There exist a finite, non-empty set P (subsequently to be interpreted as the set of particles), a set T containing two elements (later construed as points in time before and after the collision, respectively), a vectorial quantity v and a positive scalar quantity m (afterwards to become velocity and mass, respectively). To this purely mathematical characterization the law of the conservation of momentum is added. This law is framed in the concepts just given and is supposed to express the factual content of the element-core.

It has to be taken into account, on the other hand, that a formal structure of the kind just exemplified generally cannot determine unambiguously its domain of application. One and the same formal structure may be likewise applicable to several, substantively distinct classes of strongly complete with respect to that concept. It will turn out that this is not quite true (cf. II.3.§5) but there is enough similarity to make a closer examination worthwhile.

To get a clear idea about Sneed-theoreticity the example of classical collision mechanics is again useful. The question to be asked is, then, in what way can the law of the conservation of momentum be tested empirically. In addition, we restrict ourselves to this single theory-element and disregard its integration into a more comprehensive theory-net (namely, classical mechanics as a whole). Testing the conservation of momentum then amounts to determining the velocities and mass-values of the particles involved in a collision process and comparing the momentum-values thus obtained. As regards the velocities, this determination is supposed to unproblematic; we assume that spatial and temporal intervals can be measured without difficulty. Things are different, however, with respect to mass. Since we have confined ourselves to collision mechanics, we cannot resort to weighing the relevant particles. Weight determinations rely on forces, and the concept of force is alien to collision mechanics.

Hence the only means available is the law of momentum conservation itself. Suppose two particles with mass-values m_1 and m_2 collide and their velocities are altered by the impulse by an amount of Δv_1 and Δv_2, respectively. Then it follows from the law mentioned that the mass ratio is equal to the reciprocal ratio of the changes in the velocities: $(m_1)/(m_2) = (\Delta v_2)/(\Delta v_1)$. If in addition an arbitrary mass is chosen as unit, the pertinent mass-values are apparently fully determined.

But second thoughts emerge immediately. Since we have applied the law of the conservation of momentum for calculating particular mass-values, these values cannot be used in turn for a test of that law. After all, they have been adjusted so that the law is actually satisfied. A discrepancy between theoretical prediction and empirical outcome can obviously not occur in such a case; and if this possibility is precluded, testability is ruled out as well.

Since it was assumed that no other, independent means for determining mass-values are available, the only way to resolve this test difficulty is to draw on mass-values provided by a second application of the same law. This second application supplies us with mass-values according to the recipe just described, and these values can then be used to check conservation of momentum. This procedure presupposes, however, that the mass-values obtained are indeed correct, and the test of this presupposition makes one more application of the same law necessary which again contains an analogous presupposition, and so forth *ad infinitum*. Testing a hypothesis containing a Sneed-theoretical term thus involves an infinite regress. This constitutes Sneed's "problem of theoretical terms" (cf. Sneed 1979, 32–38; Stegmüller 1979, 17–18; Diederich 1981, 3–9; Stegmüller 1986, 33–42).

This example suggests that "mass" indeed is a theoretical term with respect to classical collision mechanics since the determination of particular mass-values is based on the central law of that theory. By contrast, "velocity" is non-theoretical with respect to that theory since velocities can be measured without invoking momentum conservation. On the other hand, the example is obviously somewhat artificial in that it deliberately ignores that mass-values can be determined by means other than the law of the conservation of momentum. Accordingly, the just-stated conclusion as to the theoreticity of mass with respect to collision mechanics does no longer hold if more realistic scenarios are taken into account. Anyway, full-fledged Sneed-theoretical terms are characterized by the feature that *every* empirical implementation is theory-dependent in the way just described. The discussion suggests, therefore, that in such cases a serious test problem surfaces.

What makes the whole story relevant for our purpose is its evident similarity to the completeness issue. For if a hypothesis is necessarily employed for evaluating one of the quantities in this same hypothesis, we have an ingredient of potential strong completeness on the hypothesis-level.[5] And in view of that similarity we are worse off than before. Whereas it appeared formerly that the test problems possibly associated with completeness were confined to the correction procedure (cf. II.1.§5), it seems now that the mere application of a concept in a strongly complete hypothesis may be thwarted by an infinite regress. It seems that empirical indicators cannot be associated with such a concept even in undistorted cases. That is, this association cannot be performed in an unambiguous and justified (i.e., non-arbitrary) fashion.

I refer to a situation in which one single hypothesis enters into every evaluation of one of the quantities figuring in it, as a *Sneed-bottleneck.*[6] A Sneed-bottleneck is a special case of potential strong completeness on the hypothesis-level if distortions are left out of consideration. And the problem is that it seems to preclude an empirical test of the pertinent hypothesis.

§ 3. Let's examine a trifle more closely the nature of the test problems that possibly arise in that context. In order to get a clearer idea of what is constitutive here, we should turn to the difficulties involved in testing theory-laden hypotheses in general. There is, first, the usual case of simple theory-ladenness (cf. II.1.§3). In that case the observation theories or hypotheses used are distinct from the explanatory hypothesis we wish to scrutinize. That these hypotheses are distinct is supposed to mean that their respective domains of application are not fully identical. Consequently, each of the observation hypotheses involved can be tested empirically without recourse to the explanatory hypothesis or to the remaining observation hypotheses figuring in the test of the explanatory hypothesis. Simple theory-ladenness thus entails a *unidirectional* dependence of the testability of the explanatory hypothesis upon the availability of observation hypotheses. The testability of these latter hypotheses may rely on some more auxiliary hypotheses which, however, can be tested without again resorting to the explanatory hypothesis itself. In sum, in simple theory-ladenness no observation hypothesis appears twice, and, consequently, no test circle emerges.

The test problems typically associated with simple theory-ladenness were pointed out early in this century by Duhem. As he argued, every hypothesis can only be confronted with experience by resorting to a host of ancillary assumptions. If a prediction derived from this system of hypotheses turns out to be empirically false, the theoretical origin of that failure cannot be localized uniquely. The unexpected outcome only shows that somewhere in the web of intertwined hypotheses that was applied to arrive at the prediction, an erroneous assumption is involved. But experience alone does not teach us which one the fault lies with. Experience alone is insufficient to distribute unambiguously praise and blame throughout this web of hypotheses. In fact, in case of a conflict between theory and evidence there always exists more than one option to repair the defect and regain agreement between theoretical prediction and experimental result. Accordingly, it is always possible to avoid the empirical refutation of any particular hypothesis by introducing adjustments and modifications in other parts of the pertinent theoretical system (cf. Duhem 1906, 183–188).

Duhem's argument does obviously not preclude that a conflict between theory and evidence arises in the first place. On the contrary, the argument precisely presupposes that a conflict can emerge, and it is concerned with the possible theoretical responses to the occurrence of counter-evidence. I refer to such a situation as a *Duhemian test-problem.* That is, a Duhemian test-problem is concerned with the difficulty of pinpointing the bearing of experimental results on individual hypotheses. Such a problem only appears if a theory (or a theoretical network) yields determinate empirical predictions, so that the accord between the various hypotheses involved can be checked.

It deserves emphasis that in weakly Einstein–Feigl complete theories only Duhemian test problems occur. That is, with respect to testability weakly complete theories are beset with the same difficulties that go along with simple theory-ladenness in general. For it is characteristic of weak completeness that the relevant observational procedures are multiply analyzable, and that there exists at least one theoretical analysis that is independent of the theory under scrutiny (cf. II.1.§3). This implies that a discrepancy between the results of these different accounts may in fact arise; their agreement is not guaranteed on logical grounds. This can easily be recognized in the analysis of the procedures for measuring current intensity (cf. II.1.§4). It is certainly logically possible that concrete values of that quantity as furnished by electromagnetic instruments turn out to be discordant with the same values as supplied by electrochemical procedures. Their accord is not ensured from the outset. And if a conflict may arise in principle, nothing but a Duhemian test problem is involved. The conclusion is that weak completeness is not afflicted with test problems that transcend those present in simple theory-ladenness. As regards strong completeness, however, matters seem to be different. Consider the case of strong completeness on the hypothesis-level. In that case one single hypothesis is used for establishing every observational indicator of one of the concepts figuring in that hypothesis. The strategy of checking the agreement between different hypotheses is obviously of no avail under these circumstances. And as the example presented in § 2 suggests, a test problem of a novel and peculiar kind appears to surface there. In a Sneed-bottleneck the occurrence of counter-evidence seems to be precluded outright. We obtain definite values for the quantity in question, to be sure, but due to an infinite regress these values cannot be independently checked.

I refer to a difficulty of this kind as a *non-Duhemian test problem.* Such a problem is characterized by the fact that as a result of an *infinite regress* in the testing procedure, no counter-evidence can in principle emerge. In fact, the underlying difficulty may equally be described as a *circularity.* If emphasis is placed on the need to perform an unending series of measurements, the problem assumes the shape of an infinite regress. If, on the other hand, the necessity of carrying out repeated measurements of the *same quantity* is stressed, the test problem has the look of a circle. As a result of the infinite regress or circularity involved in non-Duhemian test-problems, agreement between theory and evidence is guaranteed in advance. Sneed's "problem of theoretical terms" can now be reconstructed as the claim that Sneed bottlenecks lead to non-Duhemian test problems. It is clear that test problems of a non-Duhemian type are far more grave than Duhemian ones, and if Sneed-bottlenecks indeed usher in such problems we should rather dispense with completeness altogether.

§ 4. In fact, however, the situation is not that dramatic (as some structuralists have meanwhile come to realize themselves). In order to see this let's return to the evaluation of mass in classical collision mechanics that constituted the origin of the misgivings about testability. In fact, the whole problem evaporates on closer inspection. The trick is to apply the law of the conservation of momentum to several different but related cases and to take advantage of the constraints that restrict the values of a theoretical quantity. The relevant constraint here is the identity constraint for mass. It says that we have to ascribe the same mass-value to the same particle figuring in different applications of a law. This constraint expresses the conservation of mass and says, in particular, that the mass values involved do not change during the interaction.

As regards the details, remember that mass-values are inferred from the changes in particle velocities that occur as a result of the collision (see § 2). Assume that we have three particles with mass-values m_1, m_2 and m_3, respectively, and that we let them collide two by two. What we observe are the velocity changes brought about by a collision between m_1 and m_2 (unprimed velocity differences), m_2 and m_3 (primed velocity differences), and m_1 and m_3 (doubly primed velocity differences). The outcome of our three collision experiments may thus succinctly be expressed as follows.

$$\frac{m_1}{m_2} = \frac{\Delta v_2}{\Delta v_1} \qquad \frac{m_2}{m_3} = \frac{\Delta v_3'}{\Delta v_2'} \qquad \frac{m_1}{m_3} = \frac{\Delta v_3''}{\Delta v_1''}$$

By multiplying the two left-hand equations with one another and equating the results with the right-hand equation, it follows:

$$(1) \qquad \frac{\Delta v_2}{\Delta v_1} \frac{\Delta v_3'}{\Delta v_2'} = \frac{\Delta v_3''}{\Delta v_1''}$$

This result critically depends on the identity constraint since it is by virtue of this principle that the mass-values remain identical across the different applications and, consequently, drop out of the resulting equation altogether (cf. Diederich 1981, 7–8; Stegmüller 1986, 59–60).

What is noteworthy here is that Equation (1) does not contain theoretical quantities anymore. The mass-values have completely disappeared from the scheme; we are only dealing with velocity differences which can be ascertained without any recourse to momentum conservation. Consequently, the empirical examination of this equation involves no difficulties and, in particular, no infinite regress. The upshot of this scenario can thus be framed as follows. A theoretical law together with a constraint may establish a relation between non-theoretical quantities alone. In that case the infinite regress is stopped. We only need three experiments instead of an unending series (which really makes a difference).

Let's pause for a minute so as to get clear about the impact of this result. In a Sneed-bottleneck the ascription of particular values to a Sneed-theoretical concept is indeed not directly verifiable since every such test presupposes the very law in question. Yet this law (along with relevant constraints) may allow for deriving relations that do no longer contain any theoretical concepts. These relations can thus be tested without in turn presupposing the law under scrutiny. If they prove correct, they confirm the premises from which they were derived. Accordingly, this procedure constitutes a test of that law, and the positive outcome of this test in turn justifies the use of that law for calculating values of the Sneed-theoretical quantity under consideration. These latter values are thus confirmed indirectly.

So we have managed to avoid any explicit calculation of theoretical values in the first place. As it appears from the example, the whole point of the procedure is that the theoretical values drop out of the equation. The upshot is, then, that for testing a hypothesis it is not always requisite to compute expressly all quantities occurring in that hypothesis. There exist efficacious test options which do not rely on such evaluations, and the hypothesis thus confirmed can then legitimately be used for inferring particular values of the quantity under consideration.[7] I call the tenet that every test of a hypothesis demands the instantiation of each quantity in this hypothesis the "*overall instantiation requirement*," and I take the upshot of this discussion as militating against this requirement. What we have applied here is a *non-instantiative* test strategy.

This result can also be put otherwise. If a circularity or an infinite regress appears in every specific evaluation of a particular quantity, I speak of an instantiation problem. We have seen, then, that an instantiation problem does not necessarily lead into a non-Duhemian test problem. The latter is characterized by the feature that every attempt to confirm a hypothesis gets stuck in a circularity or infinite regress (see § 3). The procedure described accomplishes a reduction of a non-Duhemian test problem to a Duhemian one. We are dealing with two theoretical principles (conservation of momentum and mass), possibly along with observation theories for velocities, which entail directly testable predictions that may go awry. Consequently, the concord of this theoretical net can be checked empirically. This allows for the conclusion that the non-instantiative test strategy works with (at least some) instantiation problems. So there is at least sometimes an easy way out of the test problems associated with strong completeness. Let's now examine how far this way can take us.

3. RECIPROCAL DEPENDENCE OF QUANTITIES: REICHENBACH LOOPS AND NON-INSTANTIATIVE TESTING

§ 1. In order to explore the bearing and the limits of the non-instantiative strategy for testing strongly complete hypotheses, I apply this strategy to a slightly more complex situation. Up to now I was only concerned with the theory-dependent measurement of one single quantity. I now turn to a scenario in which one hypothesis contains *two* (or more) quantities that can only be evaluated by relying on this very hypothesis. In this case there exists a *reciprocal dependence* between these two quantities; that is, each one can only be determined if the other one is already known.

For reasons that will become clear in IV.4.§3, I refer to such a situation as a *Reichenbach-loop*. Compared to a Sneed-bottleneck, a Reichenbach-loop exhibits the further complication that even on the basis of the hypothesis in question, a unique determination of either quantity involved is ruled out. In ordinary Sneed-bottlenecks the quantity in question can be determined unambiguously if the underlying hypothesis is taken to be correct. The difficulties only arise with respect to the grounds for its acceptance. In Reichenbach-loops, by contrast, even the supposed truth of the law leaves the two quantities indeterminate. Either one of them can always be chosen arbitrarily; if the other one is properly adjusted, no conflict with experience will ever arise – or so it seems. There is an apparent circularity involved here that not only vitiates the non-arbitrary test of any particular evaluation; rather, it vitiates the reliable evaluation itself. Accordingly, a non-Duhemian test problem of an even more severe nature appears to surface here.

§ 2. In order to realize the nature and import of this peculiarity it seems best to start with an example. Suppose we wish to test the validity of Newton's law of gravitation in celestial mechanics. That is, we confine ourselves to astronomical applications of this law; and that means, to its application to the solar system. A straightforward test requires the instantiation of every quantity figuring in that law. In the case of a planet revolving around the sun this amounts to evaluating the masses involved and measuring the gravitational forces present between them. It can be seen at once that this job is hard to complete.

But we can make it more manageable by deriving from the law of gravitation, by recourse to other mechanical principles, consequence relations whose instantiation seems less intricate. The difficulty is, however, that in all consequence relations we can possibly arrive at in this way, two quantities are contained whose evaluation depends reciprocally on one another; namely, the mass of the respective central body (i.e., of the sun in the case of an orbiting planet) and the magnitude of the gravitational constant. For clarity's sake, I focus on but one such relation and simply state that the situation is similar in the other relevant cases. This relation is Kepler's third law.

We apply the law of gravitation (along with some auxiliary assumptions from Newtonian mechanics) to our planetary system and ignore the interactions between different planets (i.e., we split up the solar system in several two-body systems). From this we obtain an improved version of Kepler's third law. Let P denote the period of the respective planet and a the semimajor axis of its elliptical orbit; M_s is to express the value of the sun's mass and γ, finally, the gravitational constant.

The third law then reads:

$$(2) \qquad P^2 = \frac{4\pi^2}{\gamma M_s} a^3 \,.$$

The relevant orbital data of the planet at hand can be ascertained without invoking the law of gravitation. And since the mass of the sun appears in this relation, it seems to provide a basis for the straightforward calculation of that mass-value.

The problem, however, lies with the gravitational constant. In order to derive concrete mass-values from Equation (2), the magnitude of that constant must be known. In fact, we could use some terrestrial apparatus (such as Cavendish's torsion balance) to determine its magnitude; but we have presupposed that we are restricted to astronomical applications. Under these circumstances we are left with only one law but two unknown quantities, and this scenario obviously issues in a reciprocal dependence between these two quantities. The magnitude of the sun's mass and the value of the gravitational constant depend reciprocally upon one another; one of them must be given in order for the other one to be computable. We thus get stuck in a Reichenbach-loop.

Kepler's third law seems thus to be unfit as a serious test of the law of gravitation. But contrary to appearances, such a test is still possible. It is possible by availing ourselves of the rescue strategy of non-instantiative testing. In order to recognize this more clearly, let's introduce an alternative law of gravitation to the effect that the intensity of the gravitational force decreases inversely proportional to the distances (rather than the squared distances). Whereas the ordinary Newtonian approach entails that the mass of the sun should be proportional to $a^3 : P^2$ (as can be gleaned from Equation (2)), the alternative yields a proportionality to $a^2 : P^2$. These derivations rely on the assumption that the gravitational constant is indeed constant; only on that condition do the just given ratios depend on the sun's mass alone.

The point is that these two ratios can be ascertained without again applying the law of gravitation. Though we are still not in a position to calculate an absolute value for the mass of the sun, we can determine the relative magnitude of these ratios for several planets. Since these ratios are to be proportional to the mass of the sun, they must come out identical irrespective of the planet considered. If the orbital data of different planets lead to diverging estimates of the sun's mass, this discrepancy certainly militates against the laws entering that calculation.

This method thus offers a basis for testing the rival approaches. Only on the Newtonian account do these ratios exhibit the required independence from the planet under consideration.

This procedure bears the distinctive marks of non-instantiative testing. We are presented with a law, namely, the law of gravitation, that – in the astronomical applications under consideration – contains two quantities, namely, mass and the gravitational constant, each of whose determination depends on the other. Consequently, every attempted determination issues in a circularity. The gist of the just-sketched procedure is, then, to use the law of gravitation, along with two constraints, stipulating the constancy of the gravitational constant and the mass of the sun, respectively, so as to derive a relation that no longer contains the problematic quantities. In the ensuing relation between the semimajor axis of a planet's orbit and its period of revolution, no reference to the sun's mass and to the value of the gravitational constant is made. So, its test does not involve any particular difficulties. But in virtue of its sensitivity to the particular form of the law of gravitation, it allows us to distinguish empirically the two versions of this law.

A non-Duhemian test problem has thus been transformed into a Duhemian one by taking advantage of the non-instantiative strategy. It is possible to have empirical reasons for or against a particular law without explicitly calculating the value of every quantity occurring in that law. The non-instantiative nature of the test is even more marked in a Reichenbach-loop than in an ordinary Sneed-bottleneck. After all, even after the test has been carried out and the invariance of $a^3 : P^2$ confirmed, we are still at a loss to give particular values for either one of the quantities in question, i.e., the sun's mass and the gravitational constant. If we were indeed restricted to astronomical considerations of the kind addressed, we could simply choose a value for one of the quantities involved (we could conventionally set $\gamma = 1$, for instance). In Reichenbach-loops the dissolution of the non-Duhemian test problem does not supply at the same time a basis for the solution of the instantiation problem.

§ 3. It deserves emphasis that non-instantiative testing, worthwhile as it is, does not represent a universal remedy for the test problems associated with strong completeness. First, there exist alternative versions of the law of gravitation for which the strategy outlined is of no use, since the alternative at issue may imply the same relations among the relevant non-theoretical quantities as the original one. Consider a law in which the gravitational attraction is supposed to vary with the square of the source mass (i.e., the sun). This law entails the same period-axis ratios as the Newtonian one. These ratios thus fail to discriminate empirically between these two alternatives and, accordingly, cannot be used for testing these alternatives.

Second, there is a further complication lurking in the above use of Kepler's third law as a test of the law of gravitation. For the strategy applied made essential use of the alleged fact that the planets' semimajor axes are observable without recourse to the law of gravitation. But in reality this assumption is somewhat shaky. The semimajor axis is the average of the planet's minimum and maximum distance from the center of gravity of the solar system. It is this center that is actually at the focus of the planet's elliptical path. However, as a result of the gravitational action of the planets on the sun, it does not coincide with the center of the sun. But what we usually obtain from astronomical observations are the planetary positions with respect to the sun. So we are faced with the task of inferring the positions of the sun and the planets with respect to their common center of gravity from their relative positions.

This is in fact an easy task – if the relevant mass-values are known. But, by hypothesis, they aren't, and now we are really in trouble. For if we operate with a semimajor axis a^* that refers to the planetary orbit with respect to the sun, we *cannot* expect the $a^{*3} : P^2$-values to be constant for all planets. The value of that ratio is actually dependent upon the mass of the planet under consideration. That is, it turns out that in reality there is one more theoretical quantity involved here. We observe a^* and we have to infer a from its value by factoring in the planetary mass m, and that means eventually, by drawing on the law of gravitation.

From an in-principle point of view, testing indeed breaks down here. This failure is due to the lack of a constraint for the planetary masses. There are three types of problematic quantities involved in applying the law of gravitation (namely, the gravitational constant, the mass of the sun and the mass of the planet under consideration), but only the two former are effectively constrained. Accordingly, we remain caught in the circle. The situation changes indeed if we adduce an auxiliary constraint for the planetary masses to the effect that these masses be very small compared to the mass of the sun. Under these circumstances we may infer from the Newtonian law of gravitation that the directly observable $a^{*3} : P^2$-ratios are approximately constant. And if it turns out that these values are in far better agreement with one another than the a^{*2} : P^2-values (whose constancy the assumption of a linearly decreasing force of gravitation demands), we still have empirical support for the Newtonian approach. Moreover, it is at least sometimes possible to establish non-circularly that this mass ratio is indeed small. For if the planet at issue is orbited by a satellite, a reasoning similar to that in § 2 yields the product of the gravitational constant with the mass of that planet. Combination with the results obtained from Equation (2) thus gives the mass ratio of planet and sun. So, if we manage to bring to bear a sufficient number of sufficiently strong constraints, non-instantiative testing proves helpful for overcoming the test problems involved in Reichenbach loops. But as the example suggests, such constraints are not always easy to come by. Let's have a closer look at this problem by discussing another instance of non-instantiative testing.

§ 4. This instance concerns the problem of testing Newton's second law, i.e., the equation of motion of classical mechanics. According to this law, the total force **F** on a body equals its mass m times its acceleration **a** ($\mathbf{F} = m\mathbf{a}$).[8]

This equation constitutes the basis for the introduction of the concepts "force" and "mass" alike; what counts as a physical force and a body's mass is essentially characterized by this law. This leads one to suspect that all the actual evaluations of these quantities as well rely essentially on its validity. One may argue, for example, that the collision method for determining mass-values also depends at bottom upon the second law since momentum conservation can be derived from that law with the help of the principle of the equality of action and reaction (Newton's third law). Momentum conservation can thus be construed as a specialization of the second law.

On the basis of arguments of this sort structuralists indeed contend that "force" and "mass" are Sneed-theoretical concepts with respect to Newtonian mechanics. That is, all extant procedures for evaluating force and mass invoke specializations of the second law (cf. Balzer/Moulines/Sneed 1987, 52–54). Let's suppose for the time being that this is in fact true (I discuss the adequacy of this claim in § 5). We have then another example of a Reichenbach-loop: Two quantities are specified by one single relation with the result that there is a reciprocal dependence between the evaluations of these quantities.

This means, neither force nor mass can be evaluated in a non-arbitrary fashion, and the second law receives the air of an empty convention. Accordingly, the question arises: Is Newton's second law testable by the non-instantiative strategy?

To fix the ideas consider the competition between the Newtonian equation of motion and the central law of late medieval impetus physics, namely, that force equals the product of mass and velocity ($\mathbf{F} = m\mathbf{v}$). In fact, this example is somewhat contrived since impetus physics had a geometric body constant instead of Newtonian mass and moreover held that the velocity corresponding to the force present was not assumed instantaneously but rather some time later. Let me nonetheless make use of this concocted version for the sake of argument. We apply the Newtonian and the impetus approach to the motion that some balls receive by contracted coil springs. That is, we consider balls accelerated by contracted and subsequently released coil springs on a smooth, horizontal surface. We wish to predict the velocity the balls gain by the expansion of the springs. For this purpose we make use of the equation of motion and apply Hooke's law for springs, in addition. According to Hooke's law, the force exerted by a spring equals the spring's contraction x multiplied by a constant k that is specific for the spring employed ($F = kx$). On the Newtonian approach one obtains $ma = kx$, and by integration this yields for the balls' velocity $\mathrm{v} = \sqrt{}\,(\mathrm{k/m}) \times \mathrm{x}$. The same reasoning leads to the expression $\mathrm{v} = (\mathrm{k/m}) \times \mathrm{x}$ in the case of impetus physics.

These two predictions appear to be testable without again resorting to an equation of motion. Namely, we may apply the extensivity constraint for mass. On that constraint, the mass of combined particles is equal to the sum of the mass-values of the individual particles involved (cf. II.2.§1). In particular, if two particles of equal mass are combined, the resulting mass-value is just double the mass of each particle. To ascertain equality of mass we may have recourse to an equal arm balance. This instrument is a static device and, consequently, does not rely on any equation of motion (see § 6). Its use is thus neutral and non-committal as to either one of the rivaling approaches to be examined. Considering the coil-spring experiment with some mass, for one, and with a mass of double magnitude, for another, leads to a predicted velocity ratio of $1/\sqrt{2}$ in the Newtonian case and of 1/2 in the impetus framework. Since experience bears out the former prediction the Newtonian account is confirmed empirically.[9]

It seems as if we were already done; but in fact there is a further difficulty involved here. For the test has made use of Hooke's law, and the impetus theoretician might respond to this experimental defeat by abandoning this law. The spring constant, he might argue, is actually not constant but varies with the spring's contraction. The observed velocity ratios provide the opportunity to determine its magnitude as a function of the contraction. This means, the value of that quantity is varied and adjusted so that it fits with the resulting velocities. This expedient is ruled out, however, if we restrict the experiments to constant values of spring contraction. On both competing approaches, equal contractions indicate equal forces. Since the velocity ratios are again in accordance with the Newtonian prediction, the corresponding equation of motion has now received empirical support as compared to its rival.

Systematically speaking, we have again taken advantage of non-instantiative testing. In particular, we have employed the respective equations of motion along with two constraints, an extensivity constraint for mass and an identity constraint for force, so as to derive relations that can be checked without presupposing an equation of motion. There are two points to be emphasized in this connection. First, in order to dissolve a Reichenbach-loop we need a constraint for each of the quantities involved. Mass and force are the relevant quantities here; the mass of the sun and the gravitational constant were the analogous quantities in § 2. Whereas in ordinary Sneed-bottlenecks one constraint is sufficient, two of them are required in the case of a reciprocal dependence. Conversely, if the problematic quantities outnumber the available constraints (as in § 3), non-instantiative testing fails. Second, the relevant constraints must be undisputed among the rivaling approaches we want to set apart empirically. This comes out clearly in the impetus example; an at first appealing candidate for a force constraint (namely, the constancy of k) had to be dismissed because it lacked general endorsement. In sum, it is possible to dissolve a Reichenbach-loop if a sufficient number of shared constraints are available. But it appears from the preceding considerations (and will be reemphasized in IV.4.§3 and IV.5.§5) that this condition is, unfortunately, not always fulfilled.

§ 5. After the discussion of a number of examples of non-instantiative testing it seems appropriate to make explicit and to summarize the characteristic features of this strategy. A law is tested in a non-instantiative fashion by deriving from it with the help of constraints (and possibly some additional auxiliary assumptions) a relation that no longer contains all the quantities appearing in that law. In order for this relation to qualify as a test of the underlying law, the following four conditions have to be satisfied.

(1) *Instantiated Consequence.* The relation obtained must be testable by instantiation. All the quantities figuring in it have to be empirically determinable; that is, one should not get stuck in another infinite regress or circularity.
(2) *Risk of failure.* The derivative relation must possess empirical content in the sense of Karl Popper (cf. Popper 1934, 113). It must be non-tautological, i.e., not be compatible with all possible values its quantities may assume.
(3) *Premises essentially used.* The relation has to depend essentially on the law to be tested. That is, removing the law from the scheme makes the derivation invalid.

Condition (3) is to operate as a safeguard against a logical trick. It is supposed, namely, to avoid the so-called "irrelevant conjunction problem".[10] Underlying that problem is the logical peculiarity that an irrelevant statement i may be conjoined to the premises of a deduction without thereby thwarting the validity of the inference. That is, if $a \rightarrow b$ holds true, then $a\&i \rightarrow b$ is equally valid – whatever i may be like. Condition (3) is supposed to make sure that i is not tested by positively establishing b. An item appearing vacuously in a derivation remains untested by the consequences.

(4) *Selective entailment.* The deduced relation has to be sensitive to at least some possible alterations in the law to be tested. That is, it is not only required that this relation be critically dependent upon the presence or absence of the law; it has to be dependent on its precise shape as well. Some changes in the underlying law must entail changes in the resulting relation. The positive confirmation of one variant of the law in question has to imply the disconfirmation of at least one alternative variant. In other words, the deduced result must put one in a position to discriminate between some alternative versions of the underlying law.[11]

The intuition behind this condition is the same as that underlying condition (2); this time with respect to the premise, not the consequence: If a law is to receive empirical credit it must face the risk of failure. It is clear, incidentally, that condition (4) actually embraces condition (3).

If a change in a premise implies a change in the result, then this premise certainly enters into the deduction non-vacuously. All the same, I keep the two conditions separate for clarity's sake.

If these conditions are met, I maintain, we have a working test procedure for a law even if not all quantities figuring in that law are explicitly determinable. Relying on this procedure we may check a hypothesis without bringing about an instantiation of the hypothesis. We need not evaluate every quantity involved in order to ensure testability. The circularity or regress problems possibly associated with testing complete hypotheses can at least sometimes be circumvented in this fashion.

§ 6. I wish to come back briefly to Sneed-theoreticity in order to contrast completeness with that concept. Against the background of Sneed-theoreticity, completeness gains a sharper profile. Sneed-theoreticity, it will be recalled, means that all procedures for applying a concept to experience, or for assigning particular values to it, make use of the laws of one particular theory (cf. II.2.§2). I have repeatedly pointed to the similarity between Sneed-theoreticity and strong Einstein–Feigl completeness on the theory-level (as outlined in II.1.§3–4). The concept of a Sneed-bottleneck, as introduced here (cf. II.2.§2), is a stronger variant of Sneed-theoreticity. It means that there is one single law (and not merely various laws of one single theory) involved in every procedure of application or evaluation assigned to a concept. Again, there is an obvious resemblance to strong Einstein–Feigl completeness on the hypothesis-level.

As regards testability, general Sneed-theoreticity and Sneed-bottlenecks are on a completely different footing, and structuralism suffers from the fact that it does not distinguish between these two items (but rather subsumes them indiscriminately under the heading of "theoretical terms"). This weakness comes out clearly if we look into the reasons advanced for assessing "force" and "mass" to be Sneed-theoretical terms with respect to classical particle mechanics.

Consider "force" first. It is argued that every method for measuring forces presupposes Newton's second law along with a special force law (such as Hooke's law or the law of gravitation) (cf. Balzer/Moulines/Sneed 1987, 52–54). But what does this alleged "presupposition of the second law" actually amount to? Balzer–Moulines–Sneed give an example by discussing the measurement of weight (i.e., of the intensity of the gravitational force on a body) by means of a spring balance. On their own analysis, this method relies on (a) Hooke's law, (b) the equality of action and reaction, and (c) a *ceteris-paribus* clause to the effect that no additional forces be present (cf. Balzer/Moulines/Sneed 1987, 68–71). The second law is thus conspicuous by its absence.

The same holds analogously for mass measurements. Balzer–Moulines–Sneed argue that mass evaluations by means of an equal arm balance make use of rigid-body mechanics. After all, the arms of the balance are supposed to be rigid. This method thus admittedly transcends the realm of particle mechanics. But, as the argument continues, rigid-body mechanics can be reduced to particle mechanics, and so the latter is actually presupposed by this method (cf. Balzer/Moulines/Sneed 1987, 54). A more comprehensive analysis would have to take the law of gravitation into account. For why is it that the equilibrium position of an equal arm balance is regarded as indicating equality of mass? What we do in fact is to compare the two masses on the pans with a third mass, namely, the earth. Since the two masses are located at equal distance from the earth, application of the law of gravitation yields that the ratio of the gravitational forces exerted on them is as the ratio of their mass-values. Equality of force thus indicates equality of mass, and by virtue of the law of the levers the former is indicated by the equilibrium of the balance. Again, the second law appears nowhere.

In fact, however, as structuralism would have it, the second law is still involved in these measuring procedures. It plays a tacit but decisive role in the background. The special laws employed are parts of classical mechanics or are reducible to it. Structuralism construes these special laws as specializations of Newton's second law in the sense of II.2.§1. That is, these laws are interpreted as additions to the basic theory-element that are supposed to hold in a restricted domain of application. The special laws are only applied to systems to which the basic law is assumed to be applicable as well. The application of a specialization is thus derivative from the validity of the basic law (cf. Balzer/Moulines/Sneed 1987, 169). Put more concretely, the application of Hooke's law or the law of gravitation to a physical system presupposes that this system is governed by Newton's second law. This is the way in which this law enters the measuring procedures under consideration.

The structuralist account is defective in that it fails to recognize an important distinction. It does not distinguish between cases in which the second law is actually used for calculating particular values of the quantities occurring in it, on the one hand, and cases in which it is merely invoked as a license for using other laws as calculation devices, on the other. This coincides precisely with the difference between the use of a law as observation theory and its use as criterion of adequacy (cf. I.3.§1, II.1.§2). In the examples just discussed, the second law is used exclusively for the latter purpose; it does not play any role in actually inferring theoretically significant values from observed data. To be sure, there exist cases in which the second law is really used as its own observation theory. The coil-spring experiments discussed in § 4 represent a case in point. What is at issue, however, is whether this is true of all pertinent measuring procedures, and it appears now that it is not. The second law is not necessarily employed as observation theory for any quantity it contains.[12]

There is thus an important ambiguity hidden in the structuralist concept of "presupposing a theory". One may presuppose it as observation theory or as criterion of adequacy, and the crucial point is that in the latter case the test of that theory has not even remotely the air of an infinite regress. There is no doubt from the outset that unfavorable evidence may indeed arise. It is perfectly possible that mass-values obtained from an equal-arm balance and force-values resulting from the application of a spring balance might not fit into the second law. There are several hypotheses involved here, and they may clash with one another. Accordingly, all test problems possibly surfacing in such a situation are of a Duhemian nature. The structuralist contention that special test problems go along with Sneed-theoreticity in general is thus mistaken.

As a result, the above-mentioned similarity between Sneed-theoreticity and strong completeness has to be qualified as follows. Strong completeness, in contrast to Sneed-theoreticity, requires that the theory or hypothesis in question is used as its own observation theory and not merely as criterion for judging the suitability of would-be measuring procedures. The use of a theory for the latter purpose only indicates a commitment to that theory, and does not pose epistemological problems of a sort relevant in the present context. A second difference between strong completeness and Sneed-theoreticity is that the former is supposed to encompass the capacity of a theory to effect necessary corrections. I have alluded to that further aspect only briefly as yet; it will be addressed at some detail in II.5. A third difference between the two concepts concerns a terminological shift in implicit emphasis. Whereas Sneed-theoreticity is essentially a property of concepts (a concept is theoretical with respect to a theory), completeness is primarily a feature of theories (a theory is complete with respect to a concept). But it seems to me that putting emphasis on the theory is putting it rightly since it is theories, and not concepts, that make the analysis of observational procedures possible.

§ 7. There is one aspect left which I briefly wish to point to. The discussion in § 6 focused on the problem whether classical mechanics, and Newton's second law in particular, is *necessary* for evaluating forces and masses. Let me now change the perspective and turn to the question if classical mechanics is *sufficient* for an analysis of the relevant measurements. As regards force and mass, it results immediately from the discussion in this section that it is sufficient. Not the second law in isolation, to be sure, but the system of mechanical laws is suitable for ascribing particular values to forces and masses. After all, there was no non-mechanical law involved in any one of the procedures presented. Since, on the other hand, non-mechanical methods for measuring these quantities exist,[13] we may conclude that classical mechanics is complete with respect to force and mass in the sense (3) as specified in II.1.§3. That is, mechanics is sufficient but not necessary for analyzing methods for evaluating mass and force – at least if the need for corrections is ignored.

But why stop with force and mass? One might as well go a step further and apply classical mechanics to the determination of lengths, velocities and accelerations, i.e., to space-time measurements. I already mentioned the possibility of analyzing the oscillations of a simple pendulum in the framework of classical mechanics (cf. I.3.§2). More specifically, the application of the equation of motion to the simple pendulum allows for the conclusion that for small angles of displacement the pendulum performs a simple harmonic oscillation. This implies in particular that successive oscillations indicate equal temporal intervals. The question why a pendulum is suited as a clock can thus be answered on the basis of mechanical laws.

In addition, mechanics is as well sufficient for the analysis and correction of at least some perturbations. I already mentioned Newton's procedure for correcting the influences of alterations in effective gravity on the period of pendulum oscillations. Another perturbation appears for large amplitudes. If the pendulum's angle of displacement is large, its period is no longer independent of the amplitude. Accordingly, if the motion of the pendulum gradually slows down as a result of air resistance, its period changes. This distorting effect can be evaluated on the basis of mechanical laws and hence be corrected. Another option is avoiding altogether the occurrence of this distortion by giving the path of the oscillating mass a cycloidal shape. It follows from mechanics that in this device, called a cycloidal pendulum, the period is in fact independent of the amplitude however large it may be. This device thus constitutes a realization of the ideal case in which the distortion under consideration no longer occurs.

Length measurements as well can be based on the laws of mechanics. If the equality of temporal intervals is determined, equality of lengths can be specified by resorting to the law of inertia. That is, such lengths are equal which an inertially moved body passes through in equal times. This shows, on the whole, that space-time measurements can in principle be based on classical mechanics alone. Since all mechanical concepts are specifiable on the basis of spatiotemporal concepts, force and mass, we may conclude that classical mechanics is indeed complete with respect to all concepts it contains. On the other hand, mechanics is clearly not necessarily involved in every such measuring procedure – at least not as an observation theory. Atomic clocks do not necessarily rely on any mechanical law. This is why mechanics is not strongly complete. The conclusion is, accordingly, that classical mechanics is complete in the above specified sense (3) with respect to every mechanical concept.

4. THE BOOTSTRAP MODEL OF TESTING HYPOTHESES: A COMEBACK FOR OVERALL INSTANTIATION?

Einstein–Feigl completeness is concerned with the self-referential traits emerging in applying or testing hypotheses. Traits of this sort are also treated within Glymour's *bootstrap account* of confirmation which addresses problems of empirically checking or finding empirical support for theories and hypotheses. Glymour is primarily getting at what I call Duhemian test problems (cf. II.2.§3). That is, he intends to specify criteria for localizing the theoretical origins of empirical successes or failures and thus to avoid Duhem's holism of confirmation. On the latter view, every empirically borne-out result counts in favor of all the hypotheses employed in its derivation. Analogously, faced with an empirical defeat we may only conclude that somewhere in the fabric of interwoven hypotheses used for arriving at the flawed prediction, an erroneous assumption is hidden. But there is no empirical basis for singling out the mistaken element. Experience alone is insufficient for distributing praise and blame around a network of interconnected assumptions.

However, as Glymour observes, in scientific practice certain data sets are taken to be relevant for specific hypotheses and irrelevant for others – contrary to the Duhemian contention. Judging from what scientists actually do, a given set of data does not indiscriminately bear on an undissectable theoretical whole but rather allows for a more specific appraisal of particular hypotheses (cf. Glymour 1980, 133–134, 146). The question is, then, how this is possible. We need an account of testing and confirmation that enables us to assess the impact of particular pieces of evidence on particular hypotheses. This means, the task Glymour sets himself is to contribute to a solution to Duhemian test problems. And what makes his approach relevant in the present context is that in the course of his investigations Glymour hits upon the problem of testing what I call complete theories or hypotheses. I outline the essentials of Glymour's account in § 1 and apply it to a historical case in § 2 and § 3. In § 4, I reinterpret this example using the conceptual framework developed in the foregoing parts of this chapter. In § 5, I discuss some important features of the bootstrap model, and in § 6 and § 7, finally, I address the question of the comparative merits of bootstrapping and non-instantiative testing.

§ 1. On the traditional hypothetico-deductive account of testing and confirmation, a theory is tested and, in the positive case, confirmed by deriving predictions from the theory and examining them empirically. For reasons I won't go into here Glymour reverses this picture. Theoretical hypotheses are confirmed by deriving these hypotheses, or more often positive instances of them, from observation statements. The route of testing and confirmation goes from the data to the theory and not the other way around.

More specifically, a test of a hypothesis is carried out by producing positive or, as the case may be, negative instantiations of it. In an instantiation of a hypothesis all the variables appearing in that hypothesis are fixed, i.e., they have received definite values. If these values actually fit the hypothesis in question the instantiation is positive; if not, it is negative. As a rule, not all quantities figuring in a hypothesis are directly measurable. Typically, some of them have to be inferred from the observed data using additional hypotheses. These auxiliary hypotheses allow one to instantiate hypotheses framed in a vocabulary that transcends purely observational terms. Moreover, a positive instantiation counts as a successful test only on the condition that the occurrence of the theoretically anticipated result was not guaranteed in advance on logical grounds alone.

Hence, in order to confirm a hypothesis two minimal requirements have to be satisfied:

(1) Every quantity in the hypothesis is actually evaluated.
(2) A negative instantiation of the hypothesis was not ruled out beforehand.

The first condition constitutes what I have called the "overall instantiation requirement" (cf. II.2.§4). According to this condition, every test of a hypothesis necessarily presupposes that all quantities figuring in the hypothesis can explicitly and unambiguously be calculated in at least one specific application. It is not demanded, accordingly, that the pertinent quantities be computable in the whole domain of application. Partial instantiation is sufficient for testing. On the second requirement, positive instantiation must not be guaranteed from the outset. This is supposed to mean that the measuring procedures which supplied the actual value used for computing a theoretical quantity could possibly have supplied other values that would have led to a negative instantiation. Another way of stating this condition is that if every quantity in the relevant hypothesis is replaced by a function of some observable quantities using the auxiliaries, the resulting relation must not be a mathematical identity. That is, incorporating the auxiliaries into the hypothesis to be tested must not generate an identity statement. This amounts to the demand that the application of the auxiliaries must not render the hypothesis in question automatically true. Put briefly and imprecisely, it has to be possible that the experiment goes awry.[14]

The process of confirming a hypothesis thus makes use of and relies on a host of auxiliary assumptions. These auxiliaries provide the instantiations of the hypothesis to be tested. Every confirmation (or disconfirmation) correspondingly holds relative to this set of ancillary hypotheses. That is, confirmation is to be construed as a three-place relation: A hypothesis is confirmed by a set of data with respect to some hypotheses or theory. While the process of confirmation thus retains holistic aspects, its outcome does not; or at least not necessarily.

The two conditions of overall instantiation and non-exclusion of counterevidence provide a basis for localizing confirmation. More precisely, they provide a basis for eliminating certain hypotheses as being non-confirmed. If a hypothesis is not instantiated or such that it automatically matches the data, its confirmation is precluded. Suppose a theory contains an equation of the sort $E = U_1 + U_2$ in which only the quantity E is measurable or computable from the data. Under these circumstances it is impossible to confirm the hypothesis since the two quantities U_1 and U_2 remain uninstantiated. The latter cannot be evaluated definitely. This consideration implies that redundant or "deoccamized" theories fare rather poorly as regards confirmation. In a deoccamized theory a measurable or computable quantity (such as E) is consistently substituted by a combination of empirically or theoretically inaccessible quantities (such as $U_1 + U_2$).[15]

Such a maneuver amounts to introducing an untestable element into a theory which reduces the overall testability of that theory and consequently detracts from its confirmation (cf. Glymour 1980, 31–32, 118–121, 134–135, 143).

There is one feature left to describe, namely, the bootstraps. Where do the bootstraps make their appearance in Glymour's bootstrap account of confirmation? The relevant point is that the auxiliary assumptions used for instantiating a hypothesis may form part of the very theory which the hypothesis to be tested belongs to as well. One of these auxiliaries may even be identical with the latter hypothesis. In this way a theory can be tested and confirmed with respect to itself; it can lift itself by its own bootstraps. It is, of course, not requisite that all hypotheses applied originate from the same theory; but the mere possibility of, so to speak, pulling yourself up by your bootstraps is remarkable enough to give a name to the approach (cf. Glymour 1980, 111, 121–122, 140).

A bootstrap test roughly takes the following form. One theoretical quantity is evaluated in several, theoretically distinct ways. It is to be demanded, then, that the results of these distinct calculations agree with one another. In particular, we may apply different theoretical hypotheses to the same set of data, or to different such sets, so as to determine the value a particular theoretical quantity assumes in a given case. Every determination of this sort has to yield the same result. This constitutes the requirement of *concordance*, and it is this requirement on which successful bootstrapping essentially relies.[16]

§ 2. It seems helpful to illuminate Glymour's bootstrap strategy by means of an example, namely, the empirical determination of atomic weights in Dalton's atomic theory. This case has been cited by Glymour himself in favor of his approach (cf. Glymour 1980, 226–263); but he fails to make explicit the logic of the measurement procedures involved and relates these procedures only superficially to the details of his theory. So there are some lacunae left which I shall try to fill out in what follows. Furthermore, this case study may serve as an additional illustration of the virtues of non-instantiative testing as outlined in II.3.§5. For the relevant procedures are indeed afflicted with a reciprocal dependence between two theoretical quantities. So, the test problem to be treated is of a rather grave and intricate nature. I pick two such procedures for measuring atomic weights both of which in fact exhibit the distinctive properties of the bootstrap strategy. The first is Dalton's original proposal and the second is Stanislao Cannizzaro's improved approach. Since bootstrapping is committed to the overall instantiation requirement, these examples illustrate non-instantiative and instantiative testing alike.

Dalton's atomic theory was already outlined in I.3.§3. Dalton assumed that weight is among the essential properties of atoms; the atoms of every element are characterized by a constant and unique weight. I also pointed already to the apparent circularity underlying the determination of atomic weights. The problem is that there are two quantities involved, namely, atomic weights and molecular formulas, and that only one type of data was available, namely, combining weights. It is assumed, then, that the combining weight is the greater the heavier the relevant atoms are and the more atoms enter one molecule. This assumption is expressed by the following *fundamental hypothesis*: The combining weight of a substance is proportional to the product of its atomic weight and the number of its atoms contained in the molecule formed in the pertinent reaction. This number is given by the molecular formula. So the point is that we have a *reciprocal dependence* between *atomic weights* and *molecular formulas*. In other words, Dalton's fundamental hypothesis contains a Reichenbach-loop. To all appearances, this peculiarity precludes any determination of atomic weights and even vitiates the empirical check of the underlying hypothesis itself.

Yet at a second glance we are much better off. Let's begin by examining how far non-instantiative testing can take us. The pivotal point for applying this strategy is the law of multiple proportions. This law says that if there exist several compounds of two elements A and B, the combining weights of B (i.e., the weights of B reacting with a fixed weight of A) stand in a ratio of small integral numbers; and the same goes, *mutatis mutandis*, for the combining weights of A. Dalton predicted the law on the basis of his atomic theory, and it is indeed an immediate consequence of the assumed constancy of atomic weight together with the fundamental hypothesis which associated combining weights with atomic weights. This prediction was immediately verified by Dalton himself. He pointed to the fact that the combining weight of oxygen in carbon dioxide (CO_2) was in fact precisely double that weight in carbon monoxide (CO), and that the same held with respect to hydrogen in methane (CH_4) and ethylene (CH_2) (cf. Nash 1967, 39–40).

Systematically speaking, Dalton employed his fundamental hypothesis along with an identity constraint for atomic weights to the effect that the atomic weight of the same element must come out equal if different compounds of that element are taken into account. And he managed to derive a relation from these two tenets, namely, the law of multiple proportions, that referred only to combining weights and was thus free from the circularity problem mentioned. Accordingly, we have an instance of non-instantiative testing.

Moreover, this relation satisfies the conditions stipulated in II.3.§5. This is obvious as regards the first two conditions: The law of multiple proportions is itself instantiated and possesses empirical content. So I need only add a few remarks as to the essential use of Dalton's theory in the derivation of that law as well as to the selective nature of this derivation. It is required that the deduced regularity put us in a position to distinguish among different versions of the underlying theory. This is in fact possible in the present example since the multiple proportions are by no means implied by every variant of the atomic theory.

In Robert Boyle's brand of this doctrine bulk matter was conceived as consisting of a hierarchically organized structure of large-scale agglomerations of fundamental particles. The immense number of atoms involved in the formation of every chemically identifiable particle cluster allows for tiny changes in the composition of substances (cf. Boyle 1661, 524–525). Consequently, this theory in no way entails the existence of small integral ratios among the combining weights of the same substances.

The same goes with respect to a competitor of Dalton's theory, namely, Berthollet's version of affinity theory. Berthollet too advocated the large particle-cluster model according to which chemical composition may vary gradually. The definite proportions that occur in some compounds were thought to result from a complex interaction of quantity-dependent chemical binding forces with forces of cohesion and elasticity which he assumed to be quantity-dependent as well. As a consequence, combining weight ratios have nothing to do with molecule formation; rather, they arise from the superposition of the various quantity-dependent forces involved. The definite proportions that a compound may display testify to the fact that the resulting superposition of forces is strongly quantity or proportion dependent; there is one marked equilibrium state (cf. Berthollet 1803, 339–346). The existence of more than one compound of two substances only reveals, then, that there is more than one equilibrium state. Not a trace of small integral numbers is present here.

The law of multiple proportions thus uniquely follows from Dalton's atomic theory. Since the conditions of essential use and selective entailment are accordingly met, it can be concluded that the successful test of that law gives confirmatory support to this theory. Note, however, that – just as in the analogous example discussed in II.3.§2 – this solution to the test problem does not imply a solution to the instantiation problem. For we are still at a loss to specify atomic weights. Without knowledge of the molecular formulas involved we cannot rule out that a combining-weight ratio indicates some multiple of the atomic weight rather than the weight itself. This is what happens indeed in the methane-ethylene case. While the fundamental hypothesis has now received some empirical support, it is still beset with the circularity problem described.

§ 3. This is where bootstrapping enters stage. Dalton's proposal for a solution to the instantiation problem was his rule of greatest simplicity (cf. I.3.§3). If only one compound of two elements A and B is known, this rule ascribes the formula AB to the compound. The theory amended in that way actually allows for an application of the bootstrap strategy. The best way to see this is to turn to a real-life example. Consider the following reasoning of Thomas Thomson, one of Dalton's early disciples.[17]

Let's start with the formation of water. 1 gram of hydrogen reacts with 8 grams of oxygen to 9 grams of water. Since water was the only known compound of these two elements the simplicity rule yields HO as molecular formula for water. If the atomic weight of hydrogen is chosen as unit, the relative atomic weight of oxygen comes out as 8. Now consider the case of nitrous oxide (retaining the unit). There are several oxides of nitrogen, and their combining weight ratios together with the simplicity rule suggest the (correct) formula NO for nitrous oxide. Its combining weight ratios are: 7.5 grams of nitrogen react with 8 grams of oxygen to form 15.5 grams nitrous oxide. On that basis we may infer the atomic weight of nitrogen to be 7.5.

This weight has been determined here by way of a detour through the atomic weight of oxygen. We could crosscheck that result by having recourse to a direct reaction between hydrogen and nitrogen. Fortunately enough, such a reaction is actually at hand, namely, the formation of ammonia. The simplicity rule provides NH as the molecular formula, and on that basis we can predict the following combining weights: 1 gram of hydrogen should combine with 7.5 grams of nitrogen. By contrast, experiment yields that 1 gram of hydrogen actually combines with 5 grams of nitrogen. The conjunction of Dalton's fundamental hypothesis and his rule of greatest simplicity has thus led to an experimental failure.

But how to respond theoretically to this failure? Thomson's rescue maneuver consisted in suspending the simplicity rule in this case. He changed the molecular formula of ammonia to NH_2. This leads to a predicted combining weight ratio of 1/2 7.5 which is in the vicinity of the correct value 5. What we have here, in fact, is a (near) compensation of two mistakes. Thomson assumed half the atomic weight of nitrogen and nearly half (actually two thirds of) the correct proportion of hydrogen in the ammonia molecule.

Dalton's amended theory thus provides a solution to the instantiation-problem. This solution is wrong, to be sure, but this failure is immaterial from a principled point of view. The simplicity rule made an independent access to the molecular formulas possible and thus succeeded in breaking up the circle. By invoking this rule the theory entailed definite results for atomic weights, and it did so by specifying molecular formulas independently from the fundamental hypothesis. Accordingly, the Reichenbach loop dissolves; the two relevant quantities are no longer determined by only one hypothesis. Moreover, Thomson's procedure amounts to a concordance test. The amended theory is applied to two different compounds and the agreement of the respective outcomes is examined. Definite instantiation and check on concordance constitute the essential ingredients of bootstrap-testing, and so we may conclude that Thomson indeed unwittingly applied Glymour's recipe. The amended theory is left with test problems of a Duhemian kind only. But as Thomson's less than fortunate expedient demonstrates, Duhemian test problems are far from trivial.

The empirical record of the simplicity rule was thus fairly unconvincing. And in view of the fact that it was a mere rule of thumb and could hardly provide a firm ground for molecular formulas anyway, there was an urgent need for an improved procedure. Such a procedure was developed by Cannizzaro in 1858, half a century after the inception of the atomic theory. Cannizzaro shrewdly exploited Amadeo Avogadro's hypothesis, formulated in 1811, which says that all gases under equal physical conditions contain an equal number of molecules per unit volume. From this hypothesis it follows that the specific weights of gases are proportional to their molecular weights. But since the constitution of the molecules is unknown, there is still no direct access to the atomic weights themselves. In order to gain access he applied Dalton's fundamental hypothesis. On this hypothesis, the combining weight of an element forming a gasiform compound is proportional to this element's atomic weight multiplied by the number of atoms of that element in the compound molecule. This hypothesis can be used to infer the relative contribution of every element to the molecular weight from the corresponding combining weights.

Consider water vapor as an example. The molecular weight, as derived from the specific weight, comes out as 18, and the combining weight ratio of hydrogen and oxygen is 1:8. It follows that hydrogen contributes with two weight units to the molecular weight of water, and oxygen with 16 units. Knowledge of the molecular structure is unnecessary for this conclusion since the number of atoms of each element in the molecule is left open.

This reasoning is then applied to a series of different gasiform compounds of one element. The recipe just sketched allows one to determine the gravimetric share of that element in every compound molecule considered. The law of multiple proportions entails that every such share is the integral multiple of a certain minimal weight. One finds, for example, that the relative portion of hydrogen is 1 in hydrochloric acid, 2 in hydrogen gas and water, 3 in ammonia, 4 in ethylene, 6 in ethanol and so forth. The minimal weight that turns up in such a series is to be identified with the respective atomic weight, that is, 1 in the case of hydrogen. With this auxiliary assumption the number of atoms in every molecule is fixed; molecular formulas can then be read off from the combining weights using the fundamental hypothesis.[18]

From a systematic point of view we encounter the same strategy as before. The Reichenbach loop involved in the fundamental hypothesis is broken up by fixing one of the quantities involved with the help of a law independent of this hypothesis. In Cannizzaro's approach Avogadro's equal-numbers assumption takes the place of Dalton's simplicity rule. This assumption offers the opportunity to use a set of data different from the combining weights, namely, specific weights. From these data an additional theoretical quantity, namely, molecular weights, can be derived without relying on the fundamental hypothesis. And on the basis of this quantity atomic weights can be ascertained.

Accordingly, determinateness is again accomplished, and it is accomplished by taking advantage of essentially the same procedure as before. An independent access to one of the relevant theoretical quantities is established, so that a definite evaluation of this quantity is obtained. There is one difference though. Whereas Dalton's method does not employ the fundamental hypothesis in the circle-breaking process of evaluating one of the quantities figuring in this hypothesis, Cannizzaro's procedure does. The distribution of the molecular weight among the elements involved is based on this hypothesis. On the other hand, the hypothesis is used in a more innocuous form since at this stage no reference is made to the atomic weights and atomic numbers separately. Correspondingly, the inherent circularity does not actually surface.

Furthermore, a check on concordance is feasible as well. One may, for instance, determine the weight of hydrogen from the hydrogen series and the weight of chlorine from the chlorine series. This allows for a prediction of the molecular weight of hydrochloric acid which can be tested by having recourse to the observed specific weight of gasiform hydrochloric acid. The principle is the same as in Thomson's concordance check of the atomic weight of nitrogen: Atomic weights must turn out to be equal irrespective of the compound used for their determination. So we have another full-grown instance of bootstrapping, which shows, moreover, that the hypothesis to be tested may figure in the test procedure itself – just as Glymour claimed (see § 1).

§ 4. The question arising naturally at this juncture is: How can these examples be accommodated by the completeness approach? In the present subsection I reconstruct the cases described in § 3 using the concepts developed before. In fact, interpreting these cases in the framework of completeness provides important insights. For as I stressed in § 2, the underlying problem has the structure of a Reichenbach loop. On the basis of Dalton's fundamental hypothesis the evaluation of atomic weights and of molecular formulas are reciprocally dependent on one another. Consequently, every definite instantiation seems to be ruled out, so that bootstrapping – relying essentially on definite instantiation – appears to be precluded as well. As a matter of fact, however, we have somehow managed to end up with definitely instantiated atomic weights. We have successfully bootstrapped a Reichenbach loop. Hence it is worthwhile to explore in which way this remarkable achievement was reached.

As already indicated, the crucial move is to bring to bear an additional hypothesis, independent of the loop in question and suitable for establishing one of the quantities making up the loop. In the first example it is Dalton's simplicity rule that fixes the molecular formula without in turn invoking the fundamental hypothesis. Likewise, in the second example it is Avogadro's equal-numbers assumption, along with an attenuated version of the fundamental hypothesis, that is applied for the same purpose. In both examples, therefore, the critical reciprocal dependence is removed by instantiating one of the relevant quantities using a second hypothesis independent of the one at issue. I call this procedure: loop dissolution through *external fixing*.

Relying on this procedure we arrive at a definite instantiation of one of the quantities involved. The complementary one can then be instantiated by employing the critical hypothesis itself. In the Dalton–Thomson scenario this means: After the molecular formulas have been fixed externally, atomic weights can be inferred by applying the fundamental hypothesis. And conversely in the Cannizzaro scenario: After the atomic weights have been evaluated externally, the molecular formulas can be derived by means of the fundamental hypothesis.

In both cases, therefore, the latter hypothesis constitutes the *only license* for that inference. The evaluation of one of the quantities in the fundamental hypothesis is based exclusively on that hypothesis. And this means that we are faced with a *Sneed bottleneck.* Accordingly, external fixing solves the instantiation problem of a Reichenbach loop by reducing the loop to a Sneed bottleneck.

The next question is: How is the test problem inherent in such bottlenecks be handled? The test strategy appealed to in the two examples can be reconstructed as non-instantiative testing. Thomson's test procedure comes down to this: Take different compounds of one element, calculate the ratios of the respective combining weights to the number of atoms involved (as supplied by the simplicity rule), and examine if in all instances these ratios come out equal. It appears that in these ratios the critical quantity (the atomic weight) has dropped out. Analogously, in Cannizzaro's test the procedure deployed amounts to a derivation of the specific weights of gasiform compounds from the specific weights of their gasiform constituents. The critical quantities (atomic weights and molecular numbers) have disappeared from the scheme.

Conceptually speaking, these procedures are exactly like those employed in relieving the Sneed bottleneck inherent in the measurement of mass as based on momentum conservation (cf. II.2.§4). By applying a constancy constraint to the effect that the atomic weight of one and the same element is identical in all its compounds, a relation could be specified in which the relevant quantity (i.e., atomic weight or molecular formula) no longer occurs. The theory establishes a relation between magnitudes measurable without recourse to the theory. The instantiation of this consequence relation constitutes a test of the underlying bottleneck hypothesis which can then legitimately be used for calculating the quantity in question.

All in all, we are presented with the following strategy for dissolving Reichenbach loops. In the first step, one of the quantities in the loop is evaluated through external fixing. The Reichenbach loop is reduced to a Sneed bottleneck in this fashion. Accordingly, the complementary quantity can be evaluated unambiguously. In the second step, the bottleneck hypothesis employed in this evaluation is checked by means of non-instantiative testing. Hence we end up with a Reichenbach loop that is instantiated in a definite and empirically justified way.

§ 5. We are left with the task of clarifying how this account fits with the bootstrap model. Before I address this task, however, let me present some of that model's features in more general terms. To begin with, Glymours's bootstrap account constitutes an important achievement and is rightly influential. It contributes to the explication and elucidation of test and confirmation procedures in science. Actually, I have already relied on some of its characteristics (such as "test by instantiation" and "guaranteed risk of failure") in the preceding sections. Moreover, the "overall instantiation requirement" – as shared by the bootstrap model – provides the background against which non-instantiative testing gains some relevance. On the basis of the hypothetico-deductive account, which takes the borne-out consequences of a hypothesis as confirmations of this hypothesis, non-instantiative testing isn't overly exciting.

I doubt, to be sure, that Glymour has really supplied us with a solution to Duhem's test problem. For the auxiliary assumptions used for instantiating a hypothesis have to be well-tested themselves. But they can in turn only be checked by relying on further auxiliary assumptions for their instantiation and so forth. The Duhemian network of interconnected hypotheses reappears with a vengeance. In the last analysis we cannot expect to obtain an unambiguous answer as to where to locate the theoretical origin of an empirical flaw. In a way Glymour himself implicitly recognizes this unfortunate trait. For he introduces the demand for a variety of evidence as an additional requirement (i.e., as a requirement not automatically contained in the bootstrap model) which is to operate as a safeguard against Duhem-type mistakes (cf. Glymour 1980, 140). Accordingly, Glymour himself appears to consider bootstrapping incapable of protecting us efficiently from getting lost in the Duhemian fabric of hypotheses. Be that as it may, however, it merely concerns the issue whether Glymour is dead right or only nearly so.

What makes the bootstrap account particularly interesting in the present context is its relation to the completeness issue. The bootstrap aspect of the bootstrap account, namely, the claim that in testing a hypothesis one may legitimately resort to other hypotheses of the same theory, or even to this hypothesis itself, evidently resembles completeness on the theory level and on the hypothesis level, respectively. True, similarity is not identity. The lowest level of data Glymour considers is the readings of measuring instruments. Evaluations of observables are accepted as given, and attention is focused on the further use of these quantities in theoretical inferences. By contrast, I treat measured quantities as inferred and address the theoretical basis of their emergence. As a consequence of his slightly different point of view, Glymour has no use for distortions. He starts off from reliable and trustworthy empirical results so that distortions and corrections lie outside his domain of interest. Anyhow, there is enough similarity left to make a closer examination worthwhile. Given the scope of the present study, what is of real concern here is the question how well bootstrapping does with regard to the test problems that possibly accompany completeness on the hypothesis level. Put the other way around, what is to be made of non-instantiative testing against the background of the bootstrap model?

The overall instantiation requirement that lies at the heart of the bootstrap account brings out its markedly anti-hypothetico-deductive leaning. On the hypothetico-deductive model, a hypothesis is confirmed by appropriately deducing some accurate observation statements from it. Hypothetico-deductive confirmation thus relies on the so-called "converse consequence condition": If a hypothesis (or a set of hypotheses) implies a correct empirical result, the hypothesis (or the pertinent set) is thereby confirmed. That is, if the outcome of a theoretical derivation turns out to be correct, the theoretical premises essentially used in this derivation receive confirmatory support.[19]

As a result of his rejection of the hypothetico-deductive account, Glymour consistently abandons the converse consequence condition (cf. Glymour 1980, 168, 287).

At first glance, abandoning converse consequence seems patently absurd, since it appears to be pervasively used in scientific practice. Consider once more the case of Kepler's laws. The derivation of an improved version of Kepler's laws from the application of Newtonian gravitational theory (along with other mechanical principles) to a somewhat idealized variant of our planetary system is universally held to lend empirical support to that theory although these laws do not contain all the quantities figuring in the law of gravitation. Neither the planetary masses nor the force of gravity appear in the derived relations. So, the confirmation of the law of gravitation by Kepler's laws seems to violate the overall instantiation requirement and to rely instead on the converse consequence condition: Kepler's laws confirm the law of gravitation because they can be derived from it and even though they don't instantiate it. How does Glymour handle cases of this kind?

Glymour's response is that Kepler's laws do not, in fact, confirm the law of gravitation; they are nonetheless relevant for the confirmation of that law. More specifically, Kepler's laws (if conjoined to principles of Newtonian mechanics) entail the existence of an inverse square force directed from the planet to the sun. What follows from these laws is, accordingly, a particular element or a special consequence of the law of gravitation, and it is this element alone (i.e., the existence of a force of that kind) that is confirmed by Kepler's laws. It is not confirmed, in particular, that gravity is a universal interaction between all masses (which is a constitutive part of the integral law of gravitation). In sum, it is only the instantiated element that is in fact confirmed (cf. Glymour 1980, 169). The bootstrap model embraces a rejection of any form of non-instantiative testing.

§ 6. The question to be asked now is whether bootstrapping really works under all circumstances; in particular, whether it works with testing complete hypotheses. After all, the use of a hypothesis for the purpose of testing this very hypothesis is explicitly licensed by the bootstrap model, and it was precisely this problem of testing complete hypotheses that led to the development and application of non-instantiative test strategies. The first case in point is the occurrence of a Sneed-bottleneck. How does the bootstrap model accommodate such cases? Glymour only once briefly alludes to a Sneed-bottleneck situation. His example is the ideal gas law: The product of the pressure p and the volume V equals the temperature T multiplied by some undetermined constant R. That is, $pV = RT$. Furthermore, this equation is supposed to provide the only means for determining a value for R, while the remaining quantities involved may be measurable independently. What we do in such a situation, Glymour argues, is to compute R-values twice using different sets of values for the other quantities. That is, we calculate $R_1 = p_1V_1/T_1$ and $R_2 = p_2V_2/T_2$ and check if indeed $R_1 = R_2$. So we use the hypothesis to be tested in the evaluation of a quantity occurring in this hypothesis (cf. Glymour 1980, 111).

Systematically speaking, Glymour's treatment amounts to reinterpreting non-instantiative testing as application of the concordance requirement to a bootstrap test. Non-instantiative testing would latch onto the relation $p_1V_1/T_1 = p_2V_2/T_2$, as derived from the gas law along with an identity constraint for the constant R, and this relation obviously does not contain R.[20]

Glymour reverses this picture and construes the situation as allowing for a double-track computation of the constant R. That is, the same law is applied to two different data sets and agreement between the results is demanded. And this procedure constitutes a test by instantiation.

The method deployed to check the bottleneck hypothesis surfacing in the determination of atomic weights can likewise be reconstructed along these lines. Instead of bringing to bear non-instantiative testing (as I did in § 4), one may just as well construe this test as a double instantiation combined with a concordance check. The description in

§ 3 was already cast in these terms. There I spoke of the definite instantiation of atomic weights and molecular formulas; definite since the values can be checked by computing them from different data sets. The account given in § 4 thus amounts to dressing this description in non-instantiative garment.

As a matter of fact, the testing of Sneed bottlenecks can always be interpreted both ways. As I mentioned in II.2.§4, in a Sneed-bottleneck the solution to the test problem indeed provides a solution to the instantiation problem. For we may legitimately use a non-instantiatively confirmed hypothesis so as to calculate values of the quantities occurring in it. So, whereas a test does not rely on instantiations, its positive outcome allows for calculating instantiations. Regarding Sneed-bottlenecks, the difference between non-instantiative testing and bootstrapping thus boils down to the issue whether instantiation is derivative from confirmation (as in the non-instantiative framework) or the other way around (as in the bootstrap model). But nothing really important seems to turn on that difference.

§ 7. But what about Reichenbach loops? In such loops any definite instantiation of the relevant quantities is precluded by the reciprocal dependence between them. And with the loss of the possibility of definite instantiation, bootstrapping topples. Non-instantiative testing, by contrast, may succeed where bootstrapping fails. At least some Reichenbach loops can be tackled by deriving from them, resorting to appropriate constraints, relations which are testable without evaluating either of the problematic quantities. But even after this hypothesis has been confirmed non-instantiatively, an unambiguous determination of these quantities is still impossible (cf. II.3.§2). In contradistinction to Sneed bottlenecks, the relevant quantities cannot be evaluated here even if the hypothesis in question is assumed to be correct and consequently used for the evaluation. The successful test of the hypothesis still leaves us with the reciprocal dependence and thus with the evaluative circle. Even after the hypothesis has passed a non-instantiative test, the bootstrap model is still at a loss to bring to bear its machinery.

This does not mean, however, that in the face of a Reichenbach loop, bootstrapping doesn't work at all. Rather, Reichenbach loops can be bootstrapped through external fixing as described in § 4. But this operation gains only a cheap victory for it works by cheating, as it were. It appeals to an additional law, independent of the one in question, in order to evaluate one of the quantities involved. The reciprocal dependence vanishes and the loop ceases to be a loop anymore.

Can all Reichenbach loops be handled this way? Certainly not. But the loops left over as intractable may possibly be of a spurious nature and may simply be done away with altogether. I mentioned such a case in § 1: An empirically determinable quantity E was supposed to depend on two uninstantiated quantities U_1 and U_2 either or which was reciprocally dependent upon the other. Slicing away a loop of this kind has no detrimental effect on the theory's empirical record but a beneficial effect on the theory's well-testedness.

A more realistic example of a spurious Reichenbach loop is the reciprocal dependence between the determination of absolute temperature T and the evaluation of the Boltzmann constant k. Statistical mechanics only fixes the product kT and thus leaves the individual quantities undetermined. In order to evaluate either quantity the other one has to be known, with the result that neither one can be evaluated unambiguously. In fact, however, nothing but a scale convention is involved here. We might as well measure temperature in terms of kT which would amount to simply removing the loop at hand.

The question is, then, whether all Reichenbach loops fall under these two categories. That is, either the loop can be made to disappear by recourse to external fixing or it can be amputated altogether. The latter operation is certainly legitimate in the case of circles generated by simple "deoccamization" (as described in § 1) or circles representing mere scale conventions. Are all Reichenbach loops either externally dissolvable or spurious all along? I think not. We will come across examples of what I take to be genuine Reichenbach loops in IV.4.§3–7. It will turn out, however, that these loops not only burst the confines of bootstrap testing; they are also not amenable to non-instantiative testing. Thus as far as I can see, the whole ground of non-instantiative testing is covered by bootstrapping as well. Both strategies seem to be empirically equivalent in this sense.

Still, an assessment of their comparative merits is possible if we take the liberty of considering counterfactual cases. Remember the non-instantiative test of Dalton's fundamental hypothesis (see § 2). It can hardly be denied, I believe, that this hypothesis *is* actually confirmed by the law of multiple proportions. It seems implausible to claim that only the simplicity rule, that is, the availability of some instantiative procedure, is suited to confer support to that hypothesis. Such a claim would imply that all the early confirmation of Dalton's theory derived from this flawed rule of thumb. It is intuitively much more appealing to say that this early confirmation did not really turn on the mistaken results of that erroneous rule. True, this discussion is admittedly counterfactual because the Reichenbach loop inherent in testing Dalton's fundamental hypothesis was later dissolved satisfactorily. But if it had turned out not to be dissolvable, I maintain, the multiple proportions would still have constituted evidence in favor of that hypothesis.

It deserves notice, in addition, that Glymour's ancillary strategy of partial instantiation – as exemplified by his treatment of the confirmatory role of Kepler's laws (see § 5) – is of no avail here. The reason is that in the case of Dalton's hypothesis entailment works only one way. That is, the law of multiple proportions follows from Dalton's atomic theory, to be sure, but nothing about atoms follows from this law. Large parts of the 19th century scientific community accounted for the law in terms of "equivalent weights," i.e., without recourse to any underlying micro-entities. In the same vein, it seems quite straightforward to give credit to Kepler's third law for the empirical confirmation of the radius-period ratios derived from it; despite the fact that the critical quantities, namely, the sun's mass and the gravitational constant, remain uninstantiated (cf. II.3.§2). To be sure, this evaluative circle can be dissolved by externally fixing the gravitational constant through terrestrial measurement (cf. II.3.§2). But it seems implausible to claim that the testability of Kepler's third law hinges on Cavendish's torsion balance. Likewise, the empirical confirmation of the predicted velocity ratios among accelerated balls (cf. II.3.§4) should certainly count in favor of the Newtonian equation of motion, notwithstanding the fact that neither mass nor force are evaluated by the procedure.

If this counterfactual assessment is granted, however, non-instantiative testing scores an important point with respect to parsimony. The auxiliary means needed for handling Reichenbach loops have to be less powerful logically. Bootstrapping is compelled to resort to full-blown ancillary hypotheses; hypotheses that really allow for a definite evaluation of one of the relevant quantities. A constraint, by contrast, is insufficient for that purpose; it does not license any particular instantiation. Accordingly, the non-instantiative method has to add less in order to make Reichenbach loops testable. The external machinery brought to bear is kept at a minimum. Conversely, the loop is treated by relying to a greater extent on its internal properties. And this pleases the advocate of completeness. The reason for preferring non-instantiative testing over bootstrapping is, accordingly, that the former checks a Reichenbach loop at a lower conceptual price than does the latter and thus it is more competitive.

It should be recognized, however, that while non-instantiative testing is in conflict with bootstrapping in the literal sense as specified by Glymour's conditions, it certainly complies with bootstrapping in spirit. For the upshot of the present discussion may be summarized thus that we may legitimately use a hypothesis for computing quantities occurring in it even in cases in which this use is prohibited on Glymour's account (and it is immaterial here whether or not such cases are actually manifested in science). That is, we may be even more confident about our capacity to lift ourselves by our own bootstraps than Glymour envisaged.

5. CORRECTIVE REICHENBACH LOOPS: THE NATURE OF SELF-REFERENTIAL DISTORTIONS

§ 1. In the preceding sections I was primarily concerned with problems related to a theory's capacity to generate the links between theoretical states assumed within its framework and appropriate empirical indicators. This capacity is supposed to extend to the treatment of distorted cases as well (cf. II.1.§2–3). It is required that it be possible to evaluate a theoretical quantity even if perturbations are present. In the present section I address the problems arising from performing corrections relying on hypotheses that are potentially complete on the hypothesis level. As I mentioned in I.3.§1, there are two principal ways of coping with distortions. Either we avoid them in the first place or we neutralize theoretically their detrimental effects. Since we wish to discuss the theory-based evaluation of quantities in distorted cases, we have to focus on the second option. As it will turn out, however, we must still resort to the first option eventually. Judging from the examples to be discussed, reference to the pure or unperturbed case is inevitable in the end.

If a measurement is influenced by a distorting factor the raw-data obtained cannot be used as they stand; rather, they have to be corrected. And in a complete theory these corrections are to be effected on the basis of the theory itself. What makes the discussion of such self-corrections interesting and worthwhile in the present context is that they apparently lead to an infinite regress. I mentioned a problem of this kind in II.1.§5.

The relevant difficulty is that in order to correct unreliable evaluations of a quantity we have to take recourse to additional measurements that are unreliable in the same sense as the original ones. The emergence of an infinite regress thus seems to be a recurrent concomitant of completeness.

How do we get out of sticky situations of this kind? I begin by reverting to the problem of correcting measurements of electromagnetic quantities as outlined above (cf. II.1.§5). Suppose we wish to measure the magnitude of an electric resistance using a galvanometer. This can be accomplished by relying on Ohm's law which gives the magnitude as the ratio of the potential difference across the resistor and the intensity of the current through it. So we are left with the problem of reliably evaluating potential differences (i.e., voltages) and current intensities.

What makes this problem appear particularly troublesome is the following peculiarity. The voltmeter readings are influenced by the magnitude of that instrument's internal resistance. If we place a voltmeter in parallel to the resistor to be measured, we actually produce a drop in the relevant voltage. The same goes analogously with respect to measurements of current intensity. If we place an ammeter in series to the resistor, that instrument's internal resistance causes a decrease in the current intensity to be evaluated. The values of both relevant quantities are thus distorted by the internal resistance of the measuring device. In order to obtain reliable results we have to correct these distortions and this requires determination of the magnitude of the distorting factor. That is, we have to measure the internal resistance. But this is easier said than done for this corrective measurement is obviously liable to the very same distortion as the original one. If we use a second instrument for evaluating the internal resistance of the first one, this measurement is vitiated by the unknown value of the internal resistance of this second device, and so forth. To all appearances, we have come to a dead-end. To be sure, we can limit the practical impact of this difficulty by making the galvanometer resistance large if it is used as voltmeter and small if it is employed as ammeter, reducing in this way the magnitude of the distorting effect; yet this is evidently not solving but rather evading the correction problem we wish to tackle. Accordingly, from a principled point of view we appear to be caught in an infinite regress.

This infinite regress would be stopped if we succeeded in measuring the internal resistance of a galvanometer without having to take into account another, unknown resistance value. In that case we could

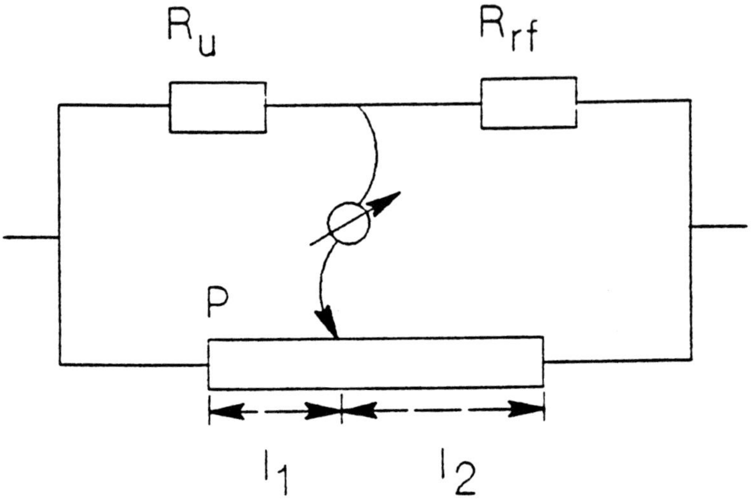

Fig. 1. Wheatstone Bridge.

reliably evaluate the magnitude of the distorting effect and thus correct it. This can indeed be achieved by making use of certain bridge networks; in particular, of the so-called Wheatstone bridge. A Wheatstone bridge is an especially arranged network of resistors placed in an electric circuit (see Fig. 1). It contains a potentiometer P (i.e., a resistor of homogeneous material equipped with a movable tap) placed in parallel to two further resistors one of which is unknown (R_u) while the other one (R_{rf}) serves as reference. R_u may represent, for instance, the sought-for internal resistance of the galvanometer. Furthermore, there is a wire connecting the two branches of the network and linking the potentiometer tap to some location between R_u and R_{rf}. This wire is connected to a galvanometer which serves the purpose of indicating if the current through the wire is actually zero. That is, the tap is moved until a zero reading of the galvanometer is obtained.

If the galvanometer actually reads zero there is no current flow between the two branches and consequently no potential difference across the connecting wire. The joint application of Kirchhoff's laws[21] and Ohm's law yields that the potential difference across R_u equals the potential difference across the potentiometer along the distance l_1; and correspondingly, the potential difference across R_{rf} is the same as the potential difference across the potentiometer along the distance l_2. Because no current flows through the connecting wire the current intensity is constant along either branch of the network. It follows that the ratio of the respective resistance values coincides in the two branches. Along with the further law that the potential drops linearly with the distance in homogeneous materials (such as the potentiometer resistor), this yields immediately: $R_u/R_{rf} = l_1/l_2$. That is, the resistor in question can be compared to some reference resistor by resorting exclusively to length measurements. And if this standard resistor is chosen as unit the value of R_u is fully determined.

The salient point of this procedure is the zero reading of the galvanometer. While it is true that this measuring instrument again possesses a non-vanishing internal resistance, the latter does not manifest itself in a measurement distortion. Since no current is actually flowing through the instrument, it follows from Ohm's law that the instrument's internal resistance causes no drop in the potential. The distorting factor is still present, to be sure, but it remains without influence. And that means, what we have realized here is the *pure or undistorted case*. We have produced a set-up in which the distortion under consideration ceases to vitiate our measurements with the result that we can now evaluate directly and reliably the internal resistance of a galvanometer. On the basis of this accomplishment we can easily rectify raw data obtained with the help of this instrument. Accordingly, the infinite regress stops after two steps. First, we determine the magnitude of the distorting factor using a set-up which is not itself subject to the distortions in question, and, second, we employ the value obtained to correct distorted measurements.

§ 2. Consider an analogous example taken from thermodynamics. Suppose we wish to determine the temperature of a cup of hot coffee by immersing a thermometer into it. The point is that the temperature of the thermometer is equal to that of the surrounding air and thus distinct from the temperature of the coffee (which is supposed to be freshly made and thus very hot). As a result, heat is transferred from the coffee to the thermometer until thermal equilibrium is reached and an equal temperature assumed. That is, by bringing thermometer and coffee into thermal contact the former is heated up and the latter is cooled off. What we obtain as measuring result is this equilibrium temperature which is different from the original coffee temperature we are after. So, the process of measurement introduces a change in the quantity to be measured.

How to cope with this distorting effect? We could attempt to perform the following correction procedure. The critical parameter for evaluating the change in temperature is the heat capacity. The heat capacity C of a system is expressed by the ratio of the amount of heat ΔQ added to (or subtracted from) the system to the amount of temperature change ΔT thereby produced ($C = \Delta Q / \Delta T$). Suppose now that the combined system of coffee and thermometer is thermally insulated. We may then conclude that the heat given off by the coffee is equal to the heat absorbed by the thermometer. In addition, the temperature increase the thermometer undergoes is directly observable. It is simply the difference between the initial and the final thermometer reading. If the relevant coffee parameters are denoted by subscripted C's and the analogous thermometer parameters by subscripted T's, the situation can be described as follows: $\Delta Q_C = -\Delta Q_T \Rightarrow C_C \Delta T_C = -C_T \Delta T_T$. So, if the heat capacities involved are given the temperature drop of the coffee as produced by the measurement can be inferred from the observable temperature increase of the thermometer. And this result can in turn be used to correct the equilibrium temperature obtained so that the coffee temperature before the immersion of the thermometer can be calculated.

Fine, we are tempted to say; but there's a fly in the ointment. For we must ask ourselves how values of the relevant heat capacities are to be obtained. And now the trouble becomes obvious since their evaluation depends in turn on temperature measurements (as it appears from the just-given description). And these temperature measurements are distorted in the same fashion as the original one we wish to correct. Correcting distorted temperature measurements thus requires recourse to temperature measurements that are subject to the very same disturbing influence. It is true that for practical purposes this difficulty does not really matter; for the magnitude of the distortion can be made small by increasing the ratio between the mass of the substance whose temperature is to be evaluated and the thermometer mass. But as I said before, circumventing the correction problem is not solving it, and from a principled point of view we seem again to be trapped in an infinite regress.

Again, there is a way out of the mess. Our problem is to evaluate reliably the magnitude of the distorting effect. This task amounts to determining the relevant heat capacities by means of temperature measurements that are not afflicted in turn with the very distortion we wish to neutralize. The trick is to start off from fixed initial values of the temperatures involved. As regards the thermometer, this can be accomplished, for instance, by making use of a heat reservoir such as the surrounding air. The thermometer reading assumed after thermal equilibrium is reached is taken as one of our starting points. It does not matter here that the temperature of the air may have been changed by the immersion of the thermometer (remember we are dealing with questions of principle). All we need for our purpose is a fixed temperature value T_{th} of the thermometer.

The second temperature value required concerns the liquid whose heat capacity we wish to ascertain. For the sake of simplicity let's take water cooled down to precisely zero degrees centigrade. This can be produced by gradually warming up ice in a vessel. When the last piece of ice has just melted the water is at the required temperature, and every additional amount of heat brings about a temperature increase. So, the water is at the known temperature T_W. Now we immerse the thermometer into the water and after a while we read the equilibrium temperature T_{eq}. Presupposing thermal insulation we may apply the above reasoning and assume that the amounts of heat absorbed and given off in this process are equal. This yields for the ratio of the heat capacities C_T and C_W of thermometer and water, respectively: $C_T/C_W = (T_{\text{eq}} - T_W)/(T_{\text{th}} - T_{\text{eq}})$. In this equation all quantities on the right-hand side can be evaluated by means of the procedure described. And if we choose the heat capacity of water as unit, the heat capacity of the thermometer is fully determined.[22]

The magnitude of the distorting heat capacity of the thermometer can thus be established empirically in this way. The actual measurement of an unknown temperature is then the precise reversal of this procedure. We use the given heat capacity of the thermometer to infer the original temperature of the water. That is, we are now in a position to actually employ the recipe sketched at the beginning of this paragraph. A transfer of this method to liquids other than water can easily be accomplished, so that the correction problem is indeed solved.

§ 3. In order to put the upshot of this discussion a bit more systematically I introduce the concept of a *self-referential distortion.* Effecting the correction of a self-referential distortion requires recourse to an evaluation of the very quantity whose measurement is to be corrected. The perturbations addressed in the two foregoing subsections are of this kind.[23]

Resistance evaluations are distorted by the internal resistance of the measuring instrument used. Evaluation (and subsequent correction) of this distortion demands recourse to a second instrument whose internal resistance has to be taken into account as well, and so forth. Internal resistance is thus a self-referential distortion of resistance measurements. The same goes analogously for the thermodynamic scenario. Measurement of temperature is distorted by the unequal temperatures of object and measuring instrument. The correction of this distortion requires reference to the heat capacities involved whose determination again relies on temperature measurements liable to the same perturbing influence. In both cases the situation has clearly all the makings of an infinite regress.

The first thing to be done now is to get the logical structure of self-referential distortions straight. In the resistance example, the correction problem stems from the fact that the resistance evaluation calls for a voltage measurement. However, this measurement yields reliable results only on the condition that the influence of the instrument's internal resistance is corrected. The same goes analogously for the relation between resistance evaluation and the measurement of current intensity. This consideration shows that the difficulty arises from a *reciprocal dependence* between resistance evaluation, on the one hand, and measurement of voltage or current intensity, on the other. And this feature arises in turn from the fact that all relevant quantities are specified by one hypothesis, namely, Ohm's law. It is this law on which the evaluation process, along with the distortion inherent in it, actually rests. We are thus presented with a case of *potential completeness on the hypothesis level*, and the trait underlying the testability problem turns out to be a *Reichenbach loop*.

Quite the same reciprocal dependence can be identified in the thermodynamic example. In order to rectify the distorted temperature reading we are in need of a reliable evaluation of the heat capacities involved. And the latter can only be obtained through a corrective procedure based on reliable temperature values.

It appears from this consideration that we are faced with *corrective*

Reichenbach loops. The definite instantiation of one of the quantities in the loop is vitiated by the distorting influence of the complementary one, and this distortion can only be evaluated by again having recourse to a measurement of the first one, and so on and so forth. Consequently, the correction of the distorting effect of the second quantity on the measurement of the first one can only be performed by making use of an evaluation of this second quantity, and this constitutes precisely the self-referential distortion pattern.

It is true that regarding the Reichenbach loops discussed in the foregoing sections this construal is rather unnatural. For it doesn't seem overly sensible to interpret the relation between the evaluation of the gravitational constant and astronomical mass measurements as a distortion of the latter by the former or vice versa; and the same goes analogously for the relation between mass and force measurements in Newtonian dynamics (cf. II.3.§2, §4). But the examples presently under discussion make it clear that there exist Reichenbach loops that lend themselves to a reconstruction as self-referential distortions.

§ 4. If we are really struggling with Reichenbach loops here we should be able to get a firmer grip on the essentials of successfully dissolving them. Is it possible, we may ask, to identify the strategies applied here with any one of those developed earlier for that purpose? In fact, it is. What we have done here amounts to loop dissolution through *external fixing* (cf. II.4.§4). In order to grasp this more clearly, let's reconsider the preceding examples from this perspective.

The essential move in blocking the infinite regress lurking in self-referential distortions was to resort to the *undistorted* case. What we have done was to employ a set-up in which the distorting factor remains mute, as it were; i.e., a set-up in which the perturbation was prevented from becoming manifest. Neither the resistance evaluation by means of the Wheatstone bridge nor the method of fixing initial values of temperature are disturbed by the effects under consideration.

How is this pure case identified? It is identified by bringing to bear additional laws, namely, Kirchhoff's laws in the resistance example. It is these laws which, when conjoined to Ohm's law, allow the conclusion that in the set-up under consideration the resistance evaluation is reliable as it stands. In particular, the outcome is unperturbed by the galvanometer's internal resistance. And quite analogously in the thermodynamic scenario. The correction procedure made use of thermometer readings obtained in thermal equilibrium and thus exempt from the distorting temperature alteration. At bottom, this procedure rests upon the second law of thermodynamics according to which temperature differences average out in the course of time. The second option was to rely on temperature values furnished by other laws of nature; in particular, by the existence of a fixed melting point. This way or other, the gist of the procedure is to appeal to temperature values uninfluenced by the distortion at issue and thus reliable as they stand.

The common element contained in all these strategies is to tackle the self-referential distortion circularity by turning to measurement procedures whose outcomes are not in need of any correction. In other words, their key aspect is the *instantiation* of the undistorted case. These strategies manage to get directly correct values of the relevant quantities. And they do so by applying additional laws, independent of the loop hypothesis in question, i.e., by external fixing. If the undistorted case can be realized this way, a direct (i.e., not in turn distorted) evaluation of the distorting factor is possible; which evaluation may then be used for correcting the original measurement. This is where the regress stops.

In I.3.§1, I introduced a distinction between actually correcting distortions and avoiding the necessity of corrections. The present discussion suggests that regarding self-referential distortions, we are actually restricted to the latter option. Rectifying a self-referential distortion appears to require instantiation of the undistorted case. We are forced to appeal to at least one instance exempt from the distortion at issue. The next question to be asked is whether this is necessarily so or whether it merely constitutes a contingent feature of the examples discussed. Given that self-referential distortions have the logical structure of Reichenbach loops, it should be possible to tackle the former by taking advantage of the alternative strategy for coping with the latter, namely, non-instantiative testing. And non-instantiative testing clearly does not require instantiation of the pure case. This presumption is as yet supported only by a rather general consideration. But it will turn out to be justified. We will encounter a relevant example in IV.5.§4–5.

6. CONCLUSION

In order that the reader not lose track of the essentials after following the various detours and ramifications of the preceding sections, I want to recapitulate the main lines of the argumentation. The chief concern of the foregoing two chapters was to investigate the relations between theoretical states and their empirical indicators. In general, these relations are established by *observation theories*. Such auxiliary theories offer a justification for specific state-indicator connections, and they license in particular the inference from specific values of an observational quantity (such as the readings of a measuring instrument) to the corresponding values of the associated theoretical quantity (cf. I.2.§5, I.3.§1).

In addition, observation theories often manage to perform their useful job even if impeded by unfavorable conditions, namely, by the presence of distortions. In such situations observation theories may be helpful in either guiding the identification or realization of the undistorted case or assisting the correction of the perturbed quantities. Observation theories may thus provide a basis for sifting the values of theoretical quantities out of disturbed data (cf. I.3.§1).

This means, observation theories are deployed in the theoretical interpretation and rectification of the empirical record. This job does *not* boil down to supplying *criteria* for acceptable measuring procedures. Criteria of adequacy stipulate requirements for what is to qualify as suitable observation theories, but they are not operative as observation theories themselves. The latter theories enable one to carry out a theoretical analysis of the relevant measuring procedures. Such theories thus not only license the connection between a theoretical state and its supposed empirical analog; they, furthermore, license the connection between specific values of the two magnitudes (cf. I.3.§1; see also II.1.§2). The intuition underlying the notion of Einstein–Feigl completeness is, then, that a theory operates as its own observation theory. The empirical means applied to realize a theoretical structure in experience can be treated and analyzed relying on that theory itself. A complete theory does not merely allow a derivation of regularities amenable to empirical test. Rather, it also allows a derivation of *procedures* suitable for testing these regularities empirically. A complete theory justifies on its own the transition from the theoretical structure to the pointer readings; and it does so not by *fiat* but by virtue of a derivation from its own principles (cf. II.1.§1–2).

To put this intuition into practice some differentiations have to be introduced. A theory may be complete with respect to a larger or a smaller number of its theoretical states. Or a theory may be sufficient or necessary or both for the pertinent theoretical analysis or for parts of this analysis (cf. II.1.§3). What deserves particular attention among these various possibilities is the case of one law entering necessarily the theoretical analysis of the measuring procedures tied to one of the concepts appearing in this law. This feature constitutes *strong completeness on the hypothesis-level* (cf. II.1.§4), and what is distinctive about it is that, on the face of it, it issues in a particular type of test problems, namely, *non-Duhemian test problems*.

Non-Duhemian test problems arise if in the course of the test procedure part of what is to be tested must be presupposed. Consequently, the test procedure is beset with a circularity or infinite regress that precludes the emergence of unfavorable evidence and thus vitiates any serious examination of the relation at issue (cf. II.2.§3). There are two types of situations in which such problems may possibly surface, namely, *Sneed bottlenecks* (cf. II.2.§2) and *Reichenbach loops* (cf. II.3.§1). In the former, one of the quantities figuring in a hypothesis can only be evaluated by recourse to that very hypothesis, in the latter, the same holds with respect to two (or more) quantities. Both types represent instances of potentially strong completeness on the hypothesis level.

Regarding Sneed bottlenecks, the difficulty takes the following form. In order to examine the validity of a hypothesis we are required to provide particular evaluations of all the quantities entering the hypothesis (overall instantiation requirement). But one of these quantities can only be instantiated by resorting to the very hypothesis at issue. It follows that this quantity is always instantiated so that it agrees with the hypothesis. And it seems, then, that no counterevidence can possibly emerge. Regarding Reichenbach loops we are even worse off. Such loops give rise to a *reciprocal dependence* between the two critical quantities. This entails that even by presupposing the validity of the hypothesis in question it is impossible to carry out an unambiguous evaluation of either quantity. One of them can be chosen conventionally and the other one appropriately adjusted. Again, any conflict with the empirical record appears to be ruled out in principle.

As it turns out, however, an empirical test can be performed even under such seemingly hopeless circumstances. There are two chief strategies for getting out of the woods, namely, *external fixing* and *non-instantiative testing*, both of which are likewise applicable to the two principal manifestations of potentially strong completeness on the hypothesis level, namely, Sneed bottlenecks and Reichenbach loops. Hence we are given four paths for leaving the non-Duhemian jungle.

External fixing works by appeal to an additional hypothesis distinct from the one under consideration. This second hypothesis provides an independent instantiation of one of the relevant quantities (cf. II.4.§4). The simplest conceivable option is represented by the application of this strategy to a Sneed bottleneck. I have not addressed explicitly this option before because it isn't overly exciting. It is adduced here so as to give an overview of all options available. Consider the Sneed bottleneck involved in carrying out mass measurements on the basis of the law of the conservation of momentum (cf. II.2.§2). It is realized that the test problems associated with this scenario dissolve immediately if one resorts to an additional procedure for measuring mass values, e.g., to the equal arm balance. This procedure rests upon the law of gravitation and the law of the levers and is independent of momentum conservation (cf. II.3.§6). Through the application of this further procedure the Sneed bottleneck vanishes; mass values are no longer determined by only one hypothesis. Consequently, a concordance check of the results furnished by the two procedures is possible, and this possibility indicates that the remaining test problems are merely Duhemian in kind.

The second option is non-instantiative testing. On this strategy, the overall instantiation requirement is abandoned. The relevant hypothesis is not tested by bringing about a special instance of it; rather, it is tested by deriving from it, relying on suitable constraints, a special consequence which no longer contains the problematic quantity and can thus be instantiated in a non-circular fashion. If this consequence indeed relies essentially on the hypothesis at issue, is sensitive to its concrete shape and bears the risk of empirical failure, then a test of this consequence constitutes a test of the underlying hypothesis as well – despite lack of overall instantiation. Mass can be evaluated unambiguously by applying the law of the conservation of momentum to at least three related collision processes, taking advantage of the constancy-of-mass constraint. By means of this strategy the non-Duhemian test problem is transformed into a Duhemian one (cf. II.2.§4, II.3.§5).

Both test options are likewise available for treating the second major manifestation of potential strong completeness on the hypothesis level, namely, Reichenbach loops. External fixing makes use of a further hypothesis suitable for fixing one of the quantities in the loop without appeal to the loop hypothesis itself. This quantity is thus definitely instantiated, and the instantiation of the complementary one is achieved by a subsequent application of the loop hypothesis. Accordingly, external fixing solves the instantiation problem of Reichenbach loops by reducing the loop to a Sneed bottleneck, which can then be tackled in the ways just described (cf. II.4.§3).

The last option, finally, is to apply the non-instantiative strategy to a Reichenbach loop. This option is implemented by bringing to bear appropriate constraints; i.e., constraints which license the derivation of a consequence relation from the loop hypothesis. If the consequence relation can be instantiated and possesses the further qualities mentioned above, we have a working test procedure for a Reichenbach loop. It is to be noted that, in contradistinction to a Sneed bottleneck, even the successful non-instantiative confirmation of a loop hypothesis does not supply us with instantiations of the critical quantities (cf. II.3.§2). Moreover, finding appropriate constraints is in general not an easy matter (cf. II.3.§3).

Up to now I was only concerned with establishing relations between theory and evidence under ideal conditions. Einstein–Feigl completeness, however, is supposed also to provide means for the theoretical correction of distorted observations and measurements. In addition to accounting for the particulars of the relevant measuring process, complete theories are also required to offer a theoretical basis for regaining or retaining the reliability of measurements in the presence of perturbational influences. Performing corrections in a complete fashion means basing them (in part) on the laws of the theory in whose framework the theoretical quantities in question are introduced (cf. II.1.§1).

In that connection a peculiar form of distortions shows up, namely, self-referential distortions. Rectifying a self-referential distortion demands an evaluation of the very quantity whose measurement we wish to correct. This evaluation is thus liable to the same kind of distortion as the measurement we want to rectify. It turned out that the logical pattern generating a self-referential distortion is in fact a Reichenbach loop. Self-referential distortions are manifestations of corrective Reichenbach loops (cf. II.5.§3). In the examples discussed the infinite regress surfacing in such correction procedures could only be kept at bay by identifying at least one instance in which the respective quantity appears undistorted. It was necessary to resort to a state of affairs in which the distortion is either absent or prevented from becoming effective. This strategy amounts to externally fixing the undistorted case. The latter is instantiated by recourse to additional hypotheses (cf. II.5.§4). This consideration immediately suggests that there ought to be self-referential distortions amenable to non-instantiative correction. This presumption will be borne out in IV.5.§4–5.

This discussion shows, then, that all the procedures possibly suitable for achieving completeness on the hypothesis level demand that additional principles be appealed to. The test problems that potentially complete hypotheses are afflicted with can only be overcome by taking advantage of the *integration* of these hypotheses into more comprehensive nomic networks. It is only through recourse to either constraints or independent hypotheses that the test circularity or infinite regress can be dissolved. Conversely, isolated hypotheses can never become complete; regarding the latter, the test problems are insurmountable.

Still, it makes a difference how much additional machinery we have to bring into the picture. It is this consideration which allows us to judge the comparative merits of external fixing and non-instantiative testing. The latter is distinguished by the virtue of modern cars and household appliances: saving energy. The same result (namely, a reliable test) is brought about using smaller conceptual resources. The constraints applied in non-instantiative testing lack the logical force of full-grown hypotheses, as requisite for external fixing, in that they fail to establish any definite instantiation of one of the quantities at issue. By tying together different applications of the same hypothesis, they merely restrict the range of admissible evaluations of these quantities. Constraints thus constitute a kind of coherence condition for the various applications of the potentially complete hypothesis itself. External fixing, by contrast, brings to bear another hypothesis, and this use of a more lavish conceptual apparatus is detrimental to the completeness of the hypothesis in question. Economy favors non-instantiative testing over external fixing (cf. II.4.§7). Either way, however, the test problems that may go along with strongly complete hypotheses can in principle be handled appropriately.

CHAPTER III

COMPLETENESS IN NATURAL SCIENCE AND PSYCHOLOGY

In Mahler's Seventh Symphony the dramatic opening and concluding movements embrace three intermediate movements addressing a wide variety of topics: nature's lyrical powers, a ghostly dance, and a nocturnal love serenade. This dramatic structure reflects the composer's insight that you can't keep tension high all the time. There is a need for *intermezzi* in which the dramatic pressure is relieved.

What befits a symphony should be appropriate for a book as well. So, this chapter represents my *intermezzo* movement. I present a garden variety of case studies that are intended to display more distinctly and intuitively the basic idea of completeness; the idea, that is, that a theory achieves on its own the translation of its fundamental concepts into experiential quantities. In particular, I address four such cases. The first concerns the completeness of electrodynamics with respect to the measurement of electromagnetic fields. The second section is devoted to the possible completeness of quantum mechanics which is considered the most fundamental and most comprehensive physical theory. So it comes as a surprise that quantum mechanics constitutes an example of failed completeness. And what is distinctive about this failure is, first, its strange and unparalleled nature and, second, the possibility of remedying it by adopting equally strange and unparalleled principles.

From corporeal nature to mind. In the third section I turn to the general theory of psychological measurement, i.e., psychological test theory, and explore the bearing of one of its concepts, i.e., construct validity, on the possible completeness of psychological theory. In the fourth section, finally, I address a particular psychological theory, namely, Freudian psychoanalysis. The question here is whether this theory offers complete guidance for tracking the elusive states of the unconscious. The answer will be in the affirmative.

Apart from the electrodynamics section, which is a trifle more complex, emphasis is placed exclusively on the general idea of completeness. I focus on a theory's capacity to establish empirical indicators for its theoretical concepts and thus confine myself to completeness on the theory level. That is, I do not go into completeness on the hypothesis level with the testability problems in its train, and I address the correction problem only in passing. Neither intricate logical relations nor circularity problems are discussed. Remember, this chapter is an *intermezzo*. So, the reader is invited to lean back, put up her legs, and go ahead quite relaxed.

1. ELECTRODYNAMICS, OR: HOW TO KEEP TRACK OF AN ELECTRON'S MOTION COMPLETELY

§ 1. Electrodynamics is the theory of electromagnetic interactions and thus describes the interactions between charged bodies. Its observational basis is primarily constituted by two quantities, namely, charges and their motions, that is, electric currents. The theoretical edifice, too, is erected using two chief quantities, namely, electric and magnetic field strength. The relevant interactions are described and explained in terms of these magnitudes. Moreover, the theory contains two kinds of basic equations. The first kind is concerned with the evaluation of the theoretical magnitudes on the basis of the observational ones. This inference from the data to the theoretical level is covered by Maxwell's equations. That is, Maxwell's equations treat the generation of electromagnetic fields from their sources, viz., charges and currents. In addition, these equations specify a third type of such sources, namely, electromagnetic fields themselves. These fields are in turn produced by temporally varying electric or magnetic fields.

The second kind of equations licenses the reverse mode of inference; the inference, that is, from the theoretical magnitudes to the observable interactions between the sources. This concerns the action of a given, external electric or magnetic field on charges and currents. The transition from the theoretical level to the observational one is essentially governed by one single law, namely, Lorentz's force law. Along with the four equations of Maxwell's, Lorentz's law constitutes the fifth basic law of electrodynamics. It relates the force $\mathbf{F}$ on a particle with charge q and velocity $\mathbf{v}$ to the prevailing electric and magnetic fields, $\mathbf{E}$ and $\mathbf{B}$, respectively: $\mathbf{F} = q(\mathbf{E} + \mathbf{v} \times \mathbf{B})$[1]. Accordingly, the evaluation of electromagnetic fields can be performed by having recourse to the forces exercised on charged particles by these fields. Additional reference to the Newtonian equation of motion allows one to pass over from the forces to the accelerations. In this way, Lorentz's force law relates the motions of charged particles to the presence and intensity of the electromagnetic field.

The measuring instrument that exhibits most perspicuously the direct application of this law is the oscilloscope. In this device electrons move under the combined influence of electric and magnetic fields; and their trajectory is determined by Lorentz's force law (along with the equation of motion). Consequently, the trajectory of these electrons – i.e., the position where they hit the screen – can be used for evaluating the underlying fields.

However, the evaluation of the electromagnetic field need not rely on Lorentz's law. Maxwell's equations, too, are suitable for this task. The field strength can be calculated from the prevailing charge distribution and current intensity using these equations. True, this state-indicator relation is somewhat peculiar in kind. It amounts to determining the field not by way of its effects but by way of its causes. The indicator process generates what it is supposed to measure. Anyway, on the whole there are two routes to the electromagnetic field quantities, one via Lorentz's force law and the other one via Maxwell's equations. But these routes may just as well be run through in the reverse directions. That is, it is likewise these two sets of laws that forge a link from the theoretical field quantities to observable processes.

Since both these sets obviously form part of electrodynamics, we may conclude that electrodynamics is complete on the theory level with respect to the measurement of the electromagnetic field. What type of completeness is realized here? First, I'm not aware of any possibility of measuring electromagnetic field strengths without relying on at least one of these electrodynamic laws. Electrodynamics is thus necessarily involved in every such measurement, and this is why we are dealing with strong completeness. Second, electrodynamics is not sufficient for such measurements. As mentioned above, going over from the forces on a particle to its actual trajectory demands application of at least one mechanical law, namely, the equation of motion. Analogously, the evaluation of electric currents is based on mechanical laws as well. Remember the analysis of the galvanometer's mode of operation. In addition to electrodynamics, its functioning relies on the rotational version of Hooke's law (cf. II.1.§1, §4). In sum, we are presented with completeness on the theory level in the sense (2) as specified in II.1.§3: Electrodynamics is necessary but not sufficient for attaching empirical indicators to electromagnetic fields.

§ 2. So far we were only concerned with the assignment of indicators; let's now move on to the theoretical correction of distortions. The difficulty underlying such theoretical correction procedures is that Lorentz's force law is a fair weather law, as it were. It only holds for undistorted cases. The field quantities figuring in that law refer to external fields, and their intensity is supposed to be evaluated by the motions of some test charge. But as a matter of fact these external fields are altered by the influence of the field of the moving charge itself. As a result, the test charge moves under the combined influence of the external fields and the field generated by itself, and this makes it clear that its motion does not reliably indicate the external field strength.

This means, electrodynamics highlights the existence of a perturbation intruding into the relevant measuring procedures. In general, this perturbation vitiates the reliable determination of the theoretical quantities in question. This problem is usually treated by requiring that the perturbation be small. It is demanded, accordingly, that the particle's charge be so small that the external fields are not significantly altered by its influence. This strategy amounts to avoiding the correction problem by restricting oneself to the pure case.

I wish to confront, rather than sidestep, the correction problem, and I will do so by discussing an example. The conclusion will turn out to coincide with the one drawn in II.5.§4: Appeal to the pure, i.e., undistorted, state is inevitable in the end. Consider the case of a uniform electric field in which a sphere of some dielectric material is placed. The sphere is polarized by the action of the external field, and this polarization brings about a distortion of the field. That is, the field is deformed and its strength changed in the vicinity of the sphere. We assume this distortion not to be negligible.

Our problem is to infer the undistorted field from quantities accessible within the distorted region. The distorted field results from two influences. First, the undistorted, external field that causes the polarization of the sphere's material. Second, the dielectric constant that expresses the degree to which the material at hand is susceptible to electric polarization; it expresses the response of the material to the external field. If these two quantities are given, i.e., if the external field as it prevails far off the sphere and the material's dielectric constant are known, then the strength and position dependence of the distorted field can be calculated from the laws of electrodynamics (cf. Eder 1967, 99–100).

But this was not exactly our problem. We are not getting at a derivation of the distorted parameters from undistorted data; rather, the reverse direction is at issue. So, is it possible to invert the mode of inference and to evaluate the undistorted field relying on distorted data? It is in fact. Since the undistorted field produces the distorted one, the former can be inferred from its effect. By measuring the sphere's polarization (i.e., the charge per unit area induced on its surface) and by taking into account the responsiveness of the material (i.e., the relevant dielectric constant), we are indeed in a position to calculate the intensity of the undistorted electric field; of the field, that is, as it would prevail without the presence of the perturbing factor (cf. Eder 1967, 100).

This sounds fine but there is one hitch. It concerns the problem of obtaining reliable values for the dielectric constant. This constant can be found empirically by exposing the material in question to an external field of *known* intensity and by determining the charge induced by that field. And now the ensuing difficulty becomes apparent. In order to extract the undistorted field from distorted data we have to rely on a quantity whose evaluation in turn requires reference to some other undistorted field.

So, we have come across another instance of a self-referential distortion (as specified in II.5.§3). Correcting for the distorting influence on an external field demands recourse to the reliable evaluation of this very quantity in a different application. The correction procedure invokes a magnitude (the dielectric constant) that can only be gained by applying an electric field of known intensity. The evaluation of this field is in principle subject to the same distortions as those we wish to rectify. And the moral to be drawn here is the same as in II.5.§4. In order to perform successfully the necessary correction procedure we must resort to a case in which the respective quantity appears undistorted. The correction of a self-referential distortion in one case demands recourse to another case in which no correction is needed. Again, such corrections are feasible only on the condition that the pure case can be realized and evaluated reliably.

§ 3. Up to now, the result can be stated thus that electrodynamics is to a large degree complete with respect to the measurements of the fields it introduces. There is, however, a complication involved here which becomes manifest in connection with electromagnetic radiation. Judging from electrodynamics, such radiation is constituted by rapidly changing electromagnetic fields that separate from their material sources and, by virtue of consecutive, mutual induction, propagate freely through space. The wave equation governing this process can be derived from Maxwell's equations.

So far, so good; but problems emerge when we focus on the possible detection procedures for electromagnetic radiation. One such procedure is the photoelectric effect, and the point is that this effect definitely does not obey the laws of electrodynamics. Moreover, the difficulty that gives rise to this failure is of a peculiar nature. For it was expected earlier that the photoelectric effect could be accommodated by electrodynamics. That is, the underlying process should in principle be analyzable by electrodynamic means. The drawback was that the results of such an analysis didn't match the empirically found properties of the effect.

So there is a peculiar form of incompleteness involved here. This incompleteness arises from the fact that the classical description of light breaks down if the interaction of light with matter is considered. That means, it is quantum mechanics in general and wave particle duality in particular that places restrictions on the completeness of electrodynamics. In light of these restrictions it seems worthwhile to explore whether quantum mechanics fares any better in this respect. We will, however, come to see that quantum mechanics fares even worse.

2. CAN QUANTUM-MECHANICAL DESCRIPTION OF PHYSICAL REALITY BE CONSIDERED COMPLETE?

When Albert Einstein, Boris Podolsky and Nathan Rosen posed their famous question as to the completeness of quantum mechanics, as cited in the title of this subsection, they had something different in mind. They aimed at throwing doubts on the theory's capacity to account comprehensively, i.e., in a non-statistical fashion, for the observable correlations between spatially distant quantum objects. Consequently, completeness in EPR's sense is generically different from Einstein–Feigl completeness as introduced here. Still, there are cases of fruitful ambiguity. So why not interpret their question as a challenge to examine the Einstein–Feigl completeness of quantum mechanics? As already indicated, the latter distinction is not among quantum mechanics' centers of excellence. In view of more recent theoretical and empirical results the conclusion seems inevitable that, contrary to EPR's contention, quantum mechanics is in fact complete in their sense. This suggests on the whole that quantum mechanics is much more EPR-complete than it is EF-complete. So, let's explore how the EF-incompleteness comes about.

§ 1. In order to get a clearer idea of the difficulties lurking here let's first turn to Niels Bohr's complementarity interpretation of quantum mechanics ("the Copenhagen interpretation") and afterward address John von Neumann's analysis of the measuring process ("the orthodox view"). Bohr's reasoning was roughly as follows. Quantum mechanics is characterized, among other things, by the introduction of special limitations to observation and measurement. In particular, Heisenberg's indeterminacy relations impose certain restrictions on the simultaneous measurability of a system's position and momentum, and of its energy content within a time interval, respectively. The more precise one of these quantities is measured, the more indeterminate is the conjugate one.

Underlying these relations is the fact, as Bohr argued, that every quantum measurement process involves a certain minimal interaction between the measuring instrument and the physical system under scrutiny. And this interaction exhibits the double peculiarity of being non-vanishing in principle and of being unpredictable. The very process of measurement is thus bound to produce a non-negligible and uncontrollable effect on the measuring results. The distortion of the quantum object through the process of measurement can neither be avoided nor evaluated and thus corrected. It is the upshot of the indeterminacy relations that such a corrective theory is impossible. As a consequence, there is no way to set apart empirically the undistorted behavior of the system, i.e., its pure state, from the distorting influence of the measurement. In other words, it is precluded by virtue of physical laws that we could sift out the pure state from the distorted data. In Bohr's view we should conclude from this situation that something like the pure state of a quantum system does not in fact exist. It makes no sense to speak of physical properties that no experimental set-up can ever detect. Quantum states are thus irremediably relational in kind; they express the relation between a quantum object and a measuring device.[2]

Accordingly, we are bound to be content with what the indeterminacy relations permit us to ascertain empirically. And these relations leave us with the *alternative* either to determine the spatiotemporal localization of a system or its energy-momentum characteristics. Since Bohr considered energy-momentum transfer the basis of causal explanation, we are faced with the choice of giving either a spatiotemporal or a causal description of a system and its evolution.

The characterization of a system in the classical fashion entails precisely the conjunction of these two modes of description: We specify the system's trajectory as it is caused by the energy-momentum conditions present. In quantum mechanics, consequently, we are presented with the separation of these two components of the classical description. Whereas the latter consists in the joint application of spatiotemporal and causal concepts, the quantum mechanical description only allows their *complementary* use: The two modes of description exclude one another but are still both necessary for a comprehensive elucidation of the phenomena (cf. Bohr 1928, 580–581; Bohr 1961, 309).

The non-classical manner of characterizing phenomena cannot be extended, however, to the description of set-up and outcome of an experiment. Bohr was firmly committed to the view that the classical language *must* be retained on the level of data description. He outright equated the classical description of the data with their communicability (cf. Bohr 1949, 209; Bohr 1961, 310). Whereas the substance of this claim is transparent, the reasons in its favor are somewhat opaque. Bohr's point seems to be that every measurement amounts to establishing space-time coincidences (cf. Bohr 1928, 580) and thus necessarily involves a spatiotemporal description, for one; and that every measurement implies the idea of causality in that its outcome is supposed to be caused by the object, for another (cf. Bohr 1936, 298). These two aspects appear to entail for Bohr that every experiment needs to be characterized in terms of both spatiotemporal and causal concepts.

Bohr's approach thus introduces a fundamental division between the measuring apparatus and the object under scrutiny. This division is fundamental in that different types of laws are to be invoked in order to accommodate each class of objects. The functioning of the apparatus is accounted for in the classical manner whereas the experimental object is described by recourse to quantum mechanics (cf. Bohr 1961, 310–311). Bohr's complementarity argument denies quantum mechanics access to the physical processes that connect the theoretical states it introduces with the readings of a measuring instrument. Quantum mechanics is prevented by reasons of principle from establishing any theory-indicator relation relying on its own conceptual means. This implies that quantum mechanics, if conceived in the Bohrian fashion, necessarily falls short of Einstein–Feigl completeness.

§ 2. Bohr's epistemic arguments against a quantum mechanical account of measurement failed to convince everyone. In particular, it did not hold back von Neumann from applying quantum mechanics to the measuring process. The starting point of von Neumann's approach was to conceive of the instrument as being represented by a quantum mechanical state function (ψ-function) as well.[3] A measurement thus involves a coupling between the respective state functions of object and instrument, and a theoretical analysis of the measurement consists in applying the laws of quantum mechanics to this coupled state. Let's envisage where such an approach takes us.

For the sake of simplicity we assume that the property we wish to measure can take only two values which may be dubbed "+" and "−". The state function of the object at hand (Ψ_0) is then represented by the sum of the partial state functions attached to each of these values: $\Psi_0 = c_1\psi_0(+)+c_2\psi_0(-)$.[4] The coefficients c_1 and c_2 allow a calculation of the probability with which measuring values corresponding to the respective partial states (i.e., "+" or "−") do actually turn up. Analogously, the instrument is supposed to be capable of assuming two states as well which are to be interpreted as "pointer on +" and "pointer on −". The state function of the measuring instrument (Ψ_M) is then given by: $\Psi_M = c_1'\psi_M(+) + c_2'\psi_M(-)$. The state function of the whole system (that is, of object and instrument) *before* an interaction between them takes place is the product of the two single state functions ($\Psi_{\text{whole}} = \Psi_0\Psi_M$).

As already mentioned, a measurement involves a coupling between object and instrument. *After* the two systems have been coupled quantum mechanics yields an overall state function of the following structure: $\Psi_{\text{whole}} = c_1^*\psi_0(+)\psi_M(+)+c_2^*\psi_0(-)\psi_M(-)$.[5] The terms of this expansion obviously refer to combined states of object and instrument. And the essential point is that it is impossible to break up these combined states into their constituents. This means, after the interaction has taken place the overall state function of the coupled system can no longer be factorized into a product of the state functions of object and instrument, respectively.

Consequently, after the interaction has begun it is no longer possible to attach definite properties to either object or instrument separately. Rather, the above superposition is to be viewed as a catalog of alternative conditionals: *If* the object is in the state "+", then the instrument is in the state "+" as well; *if*, however, the object is in the state "–", then the instrument is also in the state "–". It is plain that such a catalog of alternative conditionals does *not* allow one to conclude in which state the object actually is. What we obtain is a peculiar combination or "entanglement" between states of the object and states of the measuring apparatus.

Clearly, this entanglement could be removed if the state of the instrument were known. Knowledge of the latter would place us in the position to infer the state of the object. Accordingly, we could couple the instrument to another apparatus in order to determine its state. But now the mess reappears on a different level since this coupling results in another entanglement whose removal would require one more apparatus, and so on and so forth. This procedure obviously issues into an infinite regress.

Application of quantum mechanics to the analysis of the measuring process thus yields a nonsensical result: No measurement could ever come to a definite conclusion. The apparatus would forever remain in a schizophrenic state and be unable to pick one measuring value. In reality, however, the situation is quite different. What we observe is the occurrence of a definite state, namely, either "+" or "–", and not a straddling and wavering state of ambiguity (cf. Schrödinger 1935, 161–162; Jammer 1974, 474–481; Rae 1986, 53–59).

According to quantum mechanics, therefore, a physical reason for the occurrence of any measuring value does not exist. The point here is *not* that of a missing physical reason why a *particular* measuring value turns up in contrast to an alternative one. This would only amount to an indeterministic account of quantum phenomena and would not involve particularly mysterious aspects. The point rather is the missing physical reason why *any* measuring value turns up at all. In light of quantum mechanics it appears miraculous that microscopic systems display definite properties under observation. The physical interaction merely unfolds the range of possibilities but fails to induce the transition from potentiality to actuality. That experiments have definite outcomes has to be appended to the theory as an additional postulate.

This means that judging from a quantum mechanical analysis of the measuring process, quantum mechanics is again incomplete. And again, the incompleteness that shows up here is of a strange and outlandish kind. After all, in all the cases hitherto considered a theory's incompleteness was due to the need to resort to the laws of some other theory so as to account for the pertinent measuring procedures. In the quantum case, by contrast, there is no law known that governs the process of singling out a definite measuring value, i.e., the so-called "collapse of the state function." The occurrence of a definite object state after measurement is not accounted for by any known law of nature.

What appears striking in addition is that the Copenhagen interpretation and the orthodox view characterize the nature and pinpoint the origin of the quantum incompleteness in a quite different way. The former proceeds from considering the options for ascertaining those features of a physical system that are constitutive of a classical state description. In contradistinction, the orthodox view isn't bothered in the least about a classical description but is exclusively concerned with states characterized quantum mechanically. And whereas Bohr's view entails that the relevant incompleteness arises from the existence of an intractable distortion vitiating the comprehensive evaluation of the classical parameters, von Neumann's point is that the fundamental equation of quantum mechanics, i.e., Schrödinger's equation, fails to account for the collapse of the wave function. In spite of these differences, however, both approaches lead to the conclusion that quantum mechanics is incomplete in a unique and truly astonishing way.

§ 3. One may wonder whether this is really the final word on the subject or whether there is a way out of this messy quantum incompleteness. There is indeed one approach to the problem of quantum measurement which – if it proved viable – would allow the conclusion that quantum mechanics can be completed. The situation, as it emerges from von Neumann's analysis, is as follows. A quantum measurement results in a superposition of states each of which is associated with a certain value of the observable under consideration. And the difficulty is that nothing more seems to result from the measurement. In order to get any further two principles are usually adduced.

(1) One of the possible values, as given by quantum mechanics, is actually picked (collapse postulate).
(2) The coefficients c_k attached to the individual states forming the superposition are linked to the probability p for the corresponding value to obtain, i.e., $p = |c_k|^2$ (Born interpretation).

Completing quantum mechanics requires either disposing of or deriving these principles.

It constitutes the central contention of the so-called *many-worlds interpretation*, as proposed by Hugh Everett III and elaborated by John A. Wheeler, R. Neill Graham and Bryce S. DeWitt, that this goal can actually be achieved. In particular, the collapse postulate can be abandoned and the Born interpretation deduced.

The chief step in abandoning the collapse postulate consists in giving up the notion that one of the possible measuring outcomes is actually singled out. The reason that the strange collapse is theoretically intractable is, so the argument runs, that the collapse does not exist at all. There is no particular transition from potentiality to actuality. Rather, each component of the superposition is realized alike. This idea is spelled out as follows. As a result of the interaction between object and apparatus, the universe splits into equally real branches each of which corresponds to one component of the superposition. In each separate branch the measurement yields the result pertaining to the respective component. That is, in each branch a definite result is obtained though it differs, in general, from branch to branch.

This seemingly bizarre view thrives on some important theoretical assets. It is, first, capable of accommodating the undetectability of the assumed branching. After all, one may ask why no observer ever noticed this dramatic process of the continual split of the universe. There is obviously some need to explain why no one ever felt being torn apart while performing a measurement. This explanation can in fact be given. It follows from the fundamentals of the many-worlds interpretation, along with the principles of usual quantum mechanics, that the incessant branching is completely unobservable. Its observability is thus ruled out by the laws of nature – just as the dynamical detectability of inertial motion is excluded by virtue of the laws of Newtonian mechanics.

Second, the Born interpretation can be derived within this approach. The basic idea is roughly as follows. The number of branches arising from a measurement is counted using the square of the quantum mechanical coefficients c_k, viz. $|c_k|^2$, as weights. It can be shown, then, that the weighted relative frequency of branches corresponds to the probability of the associated measuring values. That is, whereas the Born interpretation stipulates that $|c_k|^2$ represents the probability of the corresponding outcome, this representation grows out of the many-worlds view as a natural consequence.[6]

These results suggest that on the many-worlds interpretation the measuring process can indeed be comprehensively analyzed by the conceptual resources of quantum mechanics. The theory seems to be sufficient, therefore, to establish indicator assignments for its own theoretical states. This interpretation thus appears to accomplish completeness for quantum mechanics.

In fact, a similar claim is entertained by the proponents of this splitting-universe model. It is argued that the "mathematical formalism of the quantum theory is capable of yielding its own interpretation" (DeWitt 1970, 33), and this distinguished interpretation is supposed to be precisely the model at issue. But this is clearly an overstatement. First, a formal algorithm, as a matter of principle, never furnishes its own interpretation. Formal structures are always multiply interpretable substantively; such a structure determines semantic content only up to isomorphism (cf. II.2.§1). Second, the many-worlds interpretation, as a matter of fact, does not follow logically from the quantum mechanical formalism. The formalism presents us with a superposition of state functions. And this superposition of state functions is *interpreted* as a "superposition" of co-existing universes (cf. Dieks 1991, 79). In addition, the Born interpretation is derived using the coefficients $|c_k|^2$ for weighting the different branches. It is thus granted beforehand that these coefficients are influential on relative frequencies, and this comes down to *interpreting* them as probabilities (cf. Ballentine 1973, 234).

Freed from its ill-woven garment, however, the argument retains a valid core. The splitting universe model manages to stick to the view that the measuring process is actually governed by the laws of quantum mechanics. There are no longer any nomological mavericks involved here. So, by accepting this model a coherent and comprehensive quantum mechanical treatment of the measuring process seems possible. And that's all that matters regarding the Einstein–Feigl completeness of quantum mechanics. Although quantum mechanics is certainly not necessary for analyzing the measuring process since the complementarity interpretation is quite possible logically and physically, it seems to be sufficient for that purpose. That is, quantum mechanics seems sufficient for connecting observational indicators with its theoretical states. The conclusion thus appears to be: If one is prepared to jettison some common-sense intuitions about the integral nature of the universe, then quantum mechanics can be considered complete in the sense (3) of the classificatory scheme given in II.1.§3: It is sufficient but not necessary for analyzing relevant measuring procedures.

But on closer inspection two problems surface that overshadow these pleasant prospects. The first problem is that a measurement is supposed to produce a multiplication of particles – each set belonging to a different universe. But Schrödinger's equation, that is assumed to govern the entire splitting process, requires conservation of mass-energy. Consequently, the many-worlds interpretation issues in a contradiction with the very dynamical law it is intended to elucidate (cf. Albert/Loewer 1988, 200–201).

The second difficulty is the so-called "democracy-of-bases" problem. It arises from the fact that the possible sets of bases for expanding a given state function staunchly adhere to an equal rights tenet, as it were. Put less figuratively, as a matter of mathematical fact, there are infinitely many ways of expanding a given state function into a superposition of basic states. None of these sets of bases is distinguished from a mathematical point of view. On the other hand, there are only a few such sets which can sensibly be interpreted physically, namely, those which are constituted by the eigenfunctions of some observable. Only in these cases can the coefficients figuring in the expansion reasonably be tied to the probabilities of observable phenomena.

These preferred sets of bases are thus actually determined by recourse to the concept of observable quantities. For this reason, the many-worlds interpretation contains an implicit reference to the notion of observation and measurement, and this notion is not incorporated into the structure of the state function. Moreover, distinguishing sets of bases on grounds of measurability runs counter to the fundamentals of the many-worlds interpretation. After all, this interpretation is inspired by the attempt to overcome the distinction between measuring apparatus and observed system and to purge quantum mechanics of every pre-theoretical notion of measurement.[7]

It is to be admitted, consequently, that even in light of the many-worlds interpretation quantum mechanics fails to pave all the way from the theoretical states to the empirical indicators. Strictly speaking, quantum mechanics is not sufficient for establishing state-indicator links. On the other hand, at least the "democracy problem" is of only minor significance if judged from the perspective of the present study. True, the theory does not determine in advance what its observational basis is like; but once we have made up our mind about what empirical states we are after, the theory gives us the guidance we need. In the framework of the splitting universe model, therefore, the quantum incompleteness can be remedied to a considerable extent. Still, the idea of an incessant splitting appears to many (including myself) too outlandish to be taken seriously.[8] If that really is the price which has to be paid for completing quantum mechanics, we had better leave it incomplete.

3. PSYCHOLOGICAL TEST THEORY: THE COMPLETE GUIDE TO YOUR NEIGHBOR'S PSYCHE?

§ 1. All the examples hitherto discussed were taken from the physical sciences. It seems appropriate to enlarge the scope and to apply the notion of completeness to psychology. This is what I'm going to do in the present section and the following one. At first sight it seems doubtful whether completeness can possibly find a home in psychology. After all, its appropriate application presupposes that reliable measurements can be carried out in the first place. However, as it is sometimes argued, all attempts to evaluate psychological quantities reliably are vitiated by a massive intrusion of distorting factors into any such procedure.

Consider introspection first. The very act of observing a psychological state of one's own alters this state and may even transform it beyond recognition.[9] Similarly, in experimental situations the knowledge that one is being observed may affect the behavior under scrutiny. Finally, the docility of patients on the Freudian couch creates pseudo-recollections which perfectly comply with the expectations of the therapist. These data are thus tinted (or outright generated) by the latter's interventions (cf. Grünbaum 1984, 242–245). In view of these perturbing effects, psychological measurements appear to be thwarted right from the start. And along with the notion of psychological measurement, the idea of deriving reliable empirical indicators of a psychological state from the laws of the corresponding psychological theory seems to come to grief as well.

Mere reference to the inevitable occurrence of distortions is, however, insufficient to buttress this claim. What one needs in addition is an argument to the effect that these distortions cannot be corrected. But I see no reason of principle against the possibility of taking into account the perturbing influence of the process of observation on the mental state to be identified, so that the mental state as it prevailed before the interference could be inferred. After all, precisely this method is employed in physical science – with the noteworthy exception of distortions described by the indeterminacy relations. But no argument has been adduced that would be apt to rule out that option in the introspective case.

Moreover, psychological observation is not irremediably infected with observer-induced perturbations. The occurrence of perturbations can in fact be avoided as psychological experimentalists do routinely. The principal strategy for achieving this is cheating. That is, the experimental subjects are misled about object and goal of the investigation. It is to be expected, accordingly, that the subjects' *relevant* psychological processes – i.e., the ones the experimentalist is really after – are unaffected by the awareness of taking part in a psychological experiment. Finally, patient memories can be (and must be) controlled by independent evidence – such as the witness of other persons. Accordingly, as Grünbaum points out, data from the couch are indeed insufficient to validate Freudian psychology; but if additional data are taken into account, state ascriptions performed on the basis of that theory are testable empirically (cf. Grünbaum 1983, 331–344). We may conclude from these preliminary considerations that psychological states are not so elusive as to defy any objective evaluation or measurement.

This being the case, I wish to explore the viability of the notion of completeness in psychology by way of discussing two cases. The first one concerns an interesting peculiarity in psychological test theory, and the second one deals with the problem of assigning unconscious mental states to persons within the framework of Freudian psychology.

§ 2. Psychological test theory specifies general criteria for the acceptability of tests, and among these criteria is the so-called *validity*. A valid test actually measures what we want it to measure. That is, high test scores indeed reflect a marked expression of the psychological trait the test purports to indicate. The general aim of a psychological test is to license inference from the behavior shown in the test to some psychological attribute. Usually, this attribute is in turn associated with (various types of) behavior, and the difficulty is that the behavior shown in taking the test is not, in general, identical to the behavior that is primarily characteristic of the relevant attribute. Consider a test that is supposed to indicate a person's readiness to take risks. This property may express itself in behavior such as reckless driving or venturing risky stock market speculations. But if we wish to evaluate this property we don't take that person to a highway or to the stock exchange. Rather, we tell him to fill out a questionnaire. Consequently, the observed behavior (that results in the test score) is quite different from the kinds of conduct primarily linked to the relevant psychological attribute. The validity of a test indicates in what measure the inference from the former to the latter is indeed justified. A valid test correctly rank-orders people according to the degree in which they exhibit the psychological trait under consideration.

The question emerging naturally at this stage is: How do we know that the test behavior is a valid indicator of the property we are after? There are some cases in which there is no real problem involved here; namely, those in which we have a neat and observable criterion for that property. Suppose we want to predict future performance in school. The criterion that any purported test of this potential has to satisfy lies readily at hand: Its scores have to correlate highly with the respective student's future school grades. But clear-cut situations of the kind are the exception rather than the rule. That is to say, in most cases no single and definite criterion for the relevant attribute is available. So, in these more difficult (and much more common) situations we have to resort to different strategies for assessing the validity of a given test. One such strategy relies on the so-called *construct validity*, and it is this property that is of particular interest in the present context.

Construct validity comes into play if a psychological property is not immediately observable. That is, if this property is not connected to a criterion that is easily accessible empirically and doubtlessly valid itself. In such cases the property is typically associated with a bundle of types of behavior, none of which is sufficient or necessary for the presence of that property. In a nutshell, a construct valid test is required to reproduce the relations between these various types of behavior as induced by the property in question. Put somewhat less enigmatically this means that if a psychological property is lawfully connected to some other psychological attributes, the scores of a test validly reflecting the property should display a tight correlation with the expression of the related attributes. Psychological theory highlights a number of other psychological states with which a given property should correlate positively, or, as the case may be, negatively, or should not correlate at all. A test of that property is construct valid if it indeed exhibits the expected correlations.

Consider the following example as given by Cronbach/Meehl. Assume we have a test X that is supposed to measure anxiety proneness. How can we validate this supposition? Assume that in applying the test we find a high correlation with sweaty palms as induced by fear of failing university exams. This finding supports our initial supposition since our (largely unarticulated and common-sense) theory of anxiety leads us to expect this correlation. On the other hand, it is still possible that in reality X measures academic aspiration in which case we would expect the very same correlation. This alternative option can be ruled out by inducing sweaty palms through economic threat or other non-academic ways of making people anxious. If we obtain an equally tight correlation in these cases, we may discard the alternative. Our original assumption is further confirmed if we find strong correlations of the relevant test scores with ratings on "tenseness" as passed by the respective subject's acquaintances and friends. The same goes if high correlations between the X scores and the amount of intellectual inefficiency as induced by a painful electric shock obtain. And this assessment is further buttressed if we find low correlations with social class or value orientation. As Cronbach/Meehl conclude from this example:

> We can have substantial confidence that X does measure anxiety proneness if the current theory of anxiety can embrace the variates which yield positive correlations, and does not predict correlations where we found none.[10]

A test is construct valid if its scores reflect the conceptual relations as established by the network of laws to which the concept in question belongs.

§ 3. Cronbach/Meehl view construct validity as an outgrowth of the theoretical context account of meaning (as described in I.2.§3). It follows from this account that making the meaning of a concept clear amounts to setting forth the network of laws in which this concept plays a role. Construct validity implements this general semantic account by requiring that every valid empirical indicator of a theoretical state should actually reflect these nomological interconnections. By virtue of these connections the concept is tied to other concepts some of which are in turn linked reliably to empirical indicators. Consequently, the network induces relations among types of behavior, and it is demanded that these relations be found empirically (cf. Cronbach/Meehl 1956, 187). In sum, if theoretical integration is constitutive of a concept's meaning, then the empirical indicators associated with the concept should satisfy the underlying theory.

In fact, it is not necessary to invoke the full-fledged context theory of meaning in order to justify construct validation. Mere allowance for multi-indicator concepts will do as well. As Carnap argued (in preparing the ground for what was later to become the double-language model as presented in I.1.§2), associating more than one empirical indicator with a concept introduces an intertwinement between defining a concept and using it in factual claims. Carnap's example is electric current intensity. This quantity can be measured, for instance, through the heat produced by the current in a resistor or through the deviation of a nearby magnetic needle. Associating both indicators with one and the same state implies a factual correlation between observable properties; namely, roughly speaking, if there is a deviation, then there is electric heat production as well, and vice versa. Definition and theory-guided application of a multi-indicator concept cannot be separated neatly (cf. Carnap 1936/37, 445) (cf. I.2.§4). And this is precisely what construct validity comes down to: The magnetic needle test is valid if its results correlate strongly with the results of the heat production test.

There are some obvious relations between construct validity and completeness. After all, the construct validity of a test is appraised in terms of the lawful connections between the psychological state in question and other relevant states. And if these connections belong to a single theory there emerges something akin to completeness on the theory-level. In that case a theoretical state is linked to an empirical indicator in view of correlations based on the corresponding theory. Consider again the example discussed. The assessment that X validly measures anxiety proneness is grounded on hypotheses that are themselves part of the theory of anxiety. In this sense the theory contributes to establishing a concept-indicator relation for one of its own concepts.

On the other hand, the concept-indicator relations in construct valid tests typically rely only in part on the theory in question. For in general that theory contributes nothing to actual test construction, i.e., to anticipating the precise shape of a valid indicator. It is not required by the notion of construct validity that the theory is suited to give substantive advice as to recommendable test items. Construct validity is only concerned with judging the adequacy of given tests; it is only applied with hindsight. And this means that in order to achieve construct validity a theory need not be used as an observation theory proper. Observation theories are characterized by the feature that they are actually used in the calculation of specific values of a theoretical quantity from observed data (cf. I.3.§1). They provide *theoretical* reasons as to why a particular device appropriately measures some theoretical state. They contribute to illuminating theoretically the mechanism of that device. Nothing of that sort is necessarily involved in construct validity; the latter may be based on mere empirical correlations without theoretical elucidation whatsoever. Consequently, the existence of a construct valid test, even if it relies on the laws of one single theory, does not entail that the corresponding theory is in fact complete.

Even so, construct validity clearly embraces completeness as a special case, and nothing more can sensibly be expected from a general account of psychological testing. It is not surprising, correspondingly, that the other features of completeness are not reflected in this account. Neither distortions and corrections nor circularities due to repeated use of one law in establishing concept-indicator relations form an essential part of construct validity. Consequently, the test problems emerging in connection with construct valid tests are typically merely Duhemian in nature. That is, they have only to do with the difficulty of pinpointing the theoretical origin of an unexpected empirical finding (cf. II.2.§3).[11] On the other hand, it would overburden a general test theory if we expected it to give a detailed account of the particulars of specific concept-indicator relations. If we want to obtain a more fine-grained image we need to turn to particular psychological theories. In the next section I discuss one such example, namely, Freudian psychoanalysis.

4. ON FREUD'S COUCH: PLUNGING COMPLETELY INTO THE UNCONSCIOUS

§ 1. The psychological theory developed by Sigmund Freud around the turn of the century constitutes the first comprehensive, detailed and empirically supported account of mental phenomena. Although it has been superseded in the mean time, it doubtlessly represents a conception that merits methodological attention. In the present section I explore the relations between Freud's general psychodynamic account – i.e., the explanatory theory – and the methods employed for imputing particular psychological states to particular people – i.e., for evaluating theoretical quantities postulated by the explanatory theory. Especially relevant are those states which are characteristic of the theory; states, that is, which everyday or folk psychology does not recognize. These are the states of the *unconscious* as introduced within Freudian psychology. My claim is that Freudian psychodynamics indeed provides methods for imputing such states, and that the theory is thus *Einstein–Feigl complete* to a considerable extent with respect to their determination.

Freud's theory starts from a principle of overall mental coherence. Apart from cases of physical brain damage, mental processes form an ordered and coherent structure. They proceed without disruption and obey intelligible rules. If a sequence of mental events appears discontinuous, incomplete or outright nonsensical, we should look for additional psychological processes that are suited to fill out these apparent lacunae. In this way we are led to postulate the existence of mental states that are apt to restore continuity and coherence. Such states or processes have to be *unconscious* for otherwise there would not seem to exist any gaps at all. By postulating these not directly observable states, Freud argued, the psychologist proceeds along the same line as the scientist who likewise introduces such states in order to reflect the uninterrupted causal interdependence of physical phenomena (cf. Freud 1900, 528–530; Freud 1940, 156–159).

The assumed psychological apparatus is characterized by a three-level structure. There is, first, the *id* which embraces innate instincts and the urges or drives of infancy. There is, second, the *ego* which harbors conscious thoughts and feelings and governs voluntary actions. Finally, the *super-ego* represents the forming influence of one's parents in early childhood and incorporates tradition and social milieu (cf. Freud 1940, 145–146). In the present context only the id and its relation to the ego are of importance.

The id is supposed to represent the original, primitive state of the mental apparatus as it is realized during infancy. It contains only instinctual, infantile or archaic elements. These elements have the form of needs and desires that demand satisfaction. The id is nothing but a collection of primitive drives continually and incessantly urging one to heed them. It never ceases to be active and is always prepared to extend its influence to the conscious zones of the mind (cf. Freud 1900, 553, 565–568). In other words, the id is constituted by a bundle of instinctual and infantile urges or *wishes* constantly demanding their *fulfillment.*

It is clear, on the other hand, that to us, well-behaved, well-socialized, well-civilized human beings that we are, most of these primitive wishes appear utterly unacceptable. Their conscious recognition would create fear. Consequently, they are prevented from becoming part of our conscious mind. Their recognition and satisfaction is blocked by a *censorship mechanism* that denies them access to the ego and pushes them back into the realm of the unconscious.

This consideration suggests an important differentiation within this latter realm. It is clear that we are not always aware of everything. A host of our memories and desires are not actually conscious at a given moment. But they are capable of becoming conscious; they are accessible at will. Such mental states are called *preconscious*. By contrast, another class of mental states is made inaccessible by an active process. The content of these states cannot be deliberately reproduced. These states are *unconscious* in the proper sense. The active process at work here is the censorship mentioned. This mechanism prevents retrieval of their content; it generates resistances against becoming aware of them. The thoughts and desires rejected by this censorship mechanism are called *repressed*. In adults (in contradistinction to infants) the id is to be identified with the collection of unconscious, repressed wishes (cf. Freud 1940, 159–163).

§ 2. This brief sketch may suffice as an account of Freud's basic psychodynamic model. The question to be addressed in the present context is: How and by means of which hypotheses do we come to know the content of these unconscious states that are unobservable by unaided inspection and introspection? In other words, what are the observation hypotheses applied for that purpose and what is their relation to the underlying explanatory premises? I claim that the former are indeed suggested to a considerable degree by the latter with the result that Freudian psychology is in large measure *complete* with respect to elucidating the content of unconscious states. Let's examine the relevant state-indicator relations more closely.

It follows immediately from the model outlined that if the censorship mechanism performs its duty regularly, there is no access to id-states. After all, that's why these states are unconscious in the first place. There is only one way to gain access to them; namely, if the censor is somewhat careless and lessens its grip on the instincts and desires surging from the unconscious. This obviously happens during sleep, and the mental activity occurring in that state is *dreaming*. So we may conclude that dreams reflect unconscious elements. Furthermore, since the id is supposed to consist of repressed instinctual and infantile wishes that are continually demanding their fulfillment, it follows that a reduced defense leads to the fulfillment, if hallucinatory, of these wishes. It is thus a *consequence* of the psychodynamic model that dreams represent the fictitious fulfillment of repressed wishes. In this way dreams open up an avenue to the repressed unconscious states; by means of dream interpretation it is possible to lift these states into the conscious and waking mind. Accordingly, dreams constitute the "royal road to a knowledge of the unconscious activity of the mind" (Freud 1900, 608; cf. *ibid.* 526).

Two corollaries follow from this account. The first one is the egoistic nature of dreams: Every dream deals with the dreamer himself.[12] It is clear that instinctual and infantile desires are directed at producing a somehow pleasant and enjoyable state of the corresponding person. More abstract or altruistic desires aren't sensibly to be attributed to the instinctual and infantile mind. Consequently, it is always the well-being of the dreamer that dreams are concerned with. Second, it is to be expected that when censorship is reerected after waking-up, it will exert its influence to push back into the unconscious the material that has escaped its vigilance during the dream. Accordingly, the censor exercises resistance against remembering the dream. As a result of this activity, the content of the dream is frequently, if not always, forgotten after waking-up (cf. Freud 1900, 517–520).

Two conclusions may be drawn from this sketchy presentation. Freud's general psychodynamic model entails, first, that the unconscious states postulated within its framework can be determined empirically by interpreting dreams; and it entails, second, that the key to this interpretation is provided by the rule that the dream content expresses the fulfillment of repressed id-wishes referring to the dreaming person.

§ 3. It is clear that these considerations cannot constitute the whole story. For the model as it stands would entail the self-abortion of every dream. The surging repressed id-wishes generate fear in the ego; after all, that's why they are repressed. If they intruded into the dream unhindered they would engender anxiety and ultimately wake up the dreamer. The fact that sleep is an enduring state shows that something has to be supplemented. More specifically, it shows that the influence of the censor is not totally eliminated during the dream but only weakened. It follows that the dream content is a *compromise* between the wishes surging from the id and the defensive action of the censorship mechanism. Accordingly, the dream expresses the relevant wish in *veiled* form. The censor seeks to hide the dream's true content and allows only a disguised version of the underlying wish to emerge. The wish identification is repressed. Accordingly, a correction procedure can be based on this mechanism. But in general, the latent content is still vastly underdetermined by the manifest content. So we need additional data in order to decipher the unconscious wish underlying the dream.

These additional data are supplied by the *free associations* of the dreamer. It follows from the basic model that we can hope to gather reliable clues to the dream's true content only from situations in which the vigilance of the censor is relaxed. And this is precisely the case in free association. The dreamer is asked to report all associations that occur to her with respect to the elements and aspects of the manifest dream. She is urged to renounce all self-criticism as to the surging thoughts and feelings and to adopt an impartial attitude. She is requested to utter candidly and unrestrictedly what is disagreeable and what appears unimportant and nonsensical. By thus reducing the censorship barrier we obtain cognitive material that is subject to the influence of the unconscious. And by relating this material to the corresponding dream elements we are in a position to determine the underlying repressed thoughts and wishes (cf. Freud 1900, 101–104; cf. also Freud 1940, 174).

Recourse to this auxiliary procedure is thus justified by the basic postulates. On the other hand, the method of free association falls short of providing a regular correction procedure. The cognitive material obtained by this method is obviously not exempt from the distorting influences of the censorship mechanism. After all, by hypothesis, the censorship is never completely turned off. The factual basis used for uncovering the latent dream content is thus contaminated by perturbations of the same general nature as the original dream material. The data the correction procedure is supposed to rely on are afflicted with distortions akin to those we wish to overcome. This scenario, to be sure, does not lead us into the intractable correction circularity of a self-referential distortion since the manifest dream content and the data arising from free association need not be distorted in precisely the same fashion. Rather, they may be contaminated in such a way that the undistorted aspects of either one could be used for the correction of the distorted aspects of the other one, respectively. But this vague option is obviously somewhat difficult to implement.

§ 5. The result is, then, that the two general strategies for coping with distortions, as outlined in I.3.§1, are of no avail here. Neither can we realize the pure or undistorted case, nor can we rectify disturbed evidence. Apart from special cases, there are no definite rules for inferring unambiguously the pure states from distorted data. So how do we deal with the distortion problem? The relevant point is that the theory provides clues as to what the pure state is like and how it is to be identified. We can judge the adequacy of particular sets of imputed unconscious states by invoking the principles of the theory.

This is in fact just what the procedure Freud advocates comes down to. In particular, he appears to invoke the following three criteria for the adequacy of dream interpretations.

(1) *The id-constraint*: The pivot of the latent content must consist of the fulfillment of a repressed instinctual or infantile wish.

Apart from the commitment to the basic psychodynamic model, there is a claim as to the *exclusiveness* of the described model of dream formation involved in this constraint. *Every* dream is centered on a repressed id-wish. Freud takes pains to make clear, to be sure, that the manifest content does not arise from such wishes exclusively. Rather, cognitive material originating from the day preceding the dream ("the day's residues") is also of importance here (cf. Freud 1900, 551; Freud 1940, 166). But it is ruled out that there exist dreams that unfold without unconscious contributions; dreams, that is, which only process remnants from waking life. Freud's argument runs as follows. During the day a lot of cognitive material is accumulated in the *pre*conscious. The psychologically significant aspects of this material, such as unsolved problems or emotion-laden events, exert a kind of cognitive pressure. They demand to be addressed and to be processed. While one is awake, this is achieved by dealing with them consciously. During sleep, however, the road to the consciousness is blocked, and so there is only one way left for the urging preconscious impulses: They are compelled to turn to the unconscious. There they excite and stir up cognitions which in turn enter the dream. This is why there are no dreams without elements originating from the id (cf. Freud 1900, 554–555).

(2) *The gestalt-criterion*: The interpreted dream must form a coherent whole. The obscure and apparently nonsensical elements of the manifest content must become intelligible and fit together.

The *gestalt*-criterion is backed by the above-mentioned principle of overall mental coherence (see § 1). If cognitive material is not consciously structured – as in dreams and free associations –, and if it all the same exhibits a clear structure on closer scrutiny, then this structure must derive from the unconscious (cf. Freud 1900, 528–529).

(3) *The retrodiction recommendation*: Some biographical event, unknown to the analyst, should be retrodicted correctly. An even more impressive variant obtains if the pertinent event had been "forgotten" (i.e., repressed) by the analysand himself.[13]

Satisfaction of this third criterion is not indispensible but constitutes a remarkable confirmation of the corresponding interpretation.

The first and the second criterion rely specifically on the psychodynamic model; the third one is analogous to the general methodological demand that theories predict "novel facts," i.e., empirical regularities that were previously unknown and unexpected. Consequently, although the pure states cannot strictly be inferred from the data using the laws of the theory, these states are still largely constrained by those laws. We have no worked-out correction procedure since there is too much room left for the ascription of unconscious states based on distorted data. Yet there exist theory-internal guidelines for restricting considerably the realm of admissible options.

On the whole, then, the psychodynamic model specifies distortions and supplies criteria for identifying the undistorted case. This is how the hidden states introduced by the theory are supposed to become accessible empirically. On the other hand, it does not, in general, provide rules suited to extract in a more or less unambiguous fashion the content of these states from distorted evidence. This is why there is no completeness with respect to the correction procedures necessary to derive unconscious states from the manifest dream content.

§ 6. It seems appropriate to present one specimen of psychoanalytic dream interpretation so that the criteria and mechanisms outlined are fleshed out a bit. Consider the "dead-nephew dream". A young girl had lived for a considerable period of time with her sister. During that period the sister's elder son had died leaving her with only one boy. One night the girl dreamed that the remaining boy lay dead in a coffin – just like the older one. In order to arrive at an interpretation of that dream some more aspects of the girl's biography need to be taken into account. While she lived with her sister a man frequently visited the house whom she fell in love with. The ensuing relationship, however, came to a sudden end, and the man stopped paying visits anymore. In the time that followed the girl took every opportunity to see him on neutral ground. The day preceding the dream she had told her therapist that she intended to go to a concert where she expected to see him once more. With these data at hand, the dream's latent content was clear to the analyst. In order to buttress his conjecture he asked the girl if she could remember that anything happened after the older child's death. She answered "at once" that on the occasion of that sad event the man, after a long time of absence, had come to her sister's house so as to offer his condolences; and she had watched him standing next to the coffin. Freud goes on to say:

> This was exactly what I had expected, and I interpreted the dream in this way: 'If now the other boy were to die, the same thing would happen. [...] You would see [the man] again under the same conditions as the other time. The dream means no more than your wish to see him once more [...]; it anticipated the glimpse you are to have of him to-day [at the concert] by a few hours' (Freud 1900, 153).

This interpretation exhibits all the characteristics of Freudian theory. Although the dream's manifest content is represented by a fearful event, something impossible to be desired, the analyst manages to unveil a hidden wish underlying the dream; the wish, namely, to see the beloved man. This wish is unacknowledged by the dreamer since her pride and code of conduct forbid her to chase after him. That is to say, the wish is repressed, and the dream represents its fulfillment. Second, the wish is erotic in origin, and that means, it is of instinctual nature. In sum, we are dealing with a repressed id-wish. Third, the wish is fulfilled in a disguised fashion. In the manifest content the beloved man is only present by implication. The displacement mechanism has worked efficaciously. Fourth, this implicit presence is unveiled by a spontaneous association of the dreamer. Fifth, Freud creates the impression that the crucial biographical event was anticipated by him. The interpretation is thus supported by a correct retrodiction and accordingly satisfies the third criterion of adequacy. Moreover, the interpreted dream turns on a repressed instinctual wish and makes sense (i.e., it forms a *gestalt*), so that the other two criteria are met here as well. In addition, the difference between the manifest content and the interpreted dream betrays the perturbing influence of the censorship mechanism whose action was uncovered using the method of free association as licensed by the psychodynamic model.

§ 7. The overall result is, then, that Freud's psychodynamic model is to a great extent, if not comprehensively, *sufficient* for *establishing* empirical indicators for the theoretical, i.e., unconscious, states introduced within its framework. Moreover, the theory is clearly *necessary* for that purpose. There exist no additional, theory-independent methods for unearthing repressed id-states. These states are peculiar to psychoanalytic theory. Since, on the one hand, some independent empirical features and regularities have to be taken into account so as to establish the relevant state-indicator links, the theory is not sufficient for that purpose. On the other hand, there is not much that has to be invoked in addition. So, the theory is certainly complete in the sense (2) of the classificatory scheme given in II.1.§3: It is necessary, but not sufficient, for setting up the relevant state-indicator connections. This implies that Freud's model is *strongly complete* with respect to determining unconscious states. Moreover, it comes close to completeness in the sense (1) of the just-mentioned scheme. That is, the theory is to a large extent sufficient for establishing empirical procedures corresponding to the posited theoretical states.

In addition, although the theory fails to specify an algorithmic correction procedure, it characterizes possible distortions and furnishes clues for identifying the undistorted states. Accordingly, there is no completeness with respect to corrections, but the theory still manages to cope with distortions by specifying criteria that all admissible pure states have to meet. The theory does not provide a basis for inferring the content of unconscious states unambiguously from distorted evidence, but it manages to restrict strongly the range of possible states of the kind.

This result shows, incidentally, that notions that have grown out of the analysis of theories from the natural sciences are applicable to psychological theory as well. It is immaterial in this connection that Freud's theory suffers from some deficits, if small, regarding completeness. Quantum mechanics is certainly worse off in this respect. So, emphasis is to be placed on applicability in principle, not on the shortcomings in detail. Accordingly, the notion of completeness affords another point of view from which the methodical unity of science becomes obvious.[14]

§ 8. In conclusion I wish to say a few words about the appropriateness of the Freudian approach to dreams. So far I was only concerned with the logical relations between theoretical postulates and associated empirical indicators, and these relations are obviously independent of the truth of the postulates. Although their truth is thus irrelevant for the present purpose (as it is the truth of most scientific theories for most of the purposes in philosophy of science), it constitutes a contentious question in some quarters. So it is appropriate to add a few sketchy remarks as to the viability of the submitted dream interpretations. Are there any good reasons for considering a Freudian dream interpretation to be an excavation rather than an implantation of unconscious states?

As I reconstruct Freud's line of reasoning, he gives two principal arguments in favor of the reliability of his results that do not presuppose the correctness of the basic model.

(1) *The gestalt argument*: It is extremely improbable that an interpretation which meets the *gestalt* criterion mentioned above and which, consequently, furnishes a coherent and exhaustive account of seemingly disordered and unrelated raw-data, represents a mere fabrication. If an interpretation succeeds in making so much sense out of apparent nonsense, it must reflect something preexistent.[15]

The trouble with this argument is that the resulting coherence may have preexisted only in the mind of the analyst. As Adolf Grünbaum points out with respect to psychopathological diagnoses (and without particular reference to dream interpretations), there is an additional distorting factor present that makes the ensuing *gestalt* appear less than convincing. This factor is the suggestive influence of the analyst. By selecting the evidence in a biased fashion and by posing leading questions the analyst may cook up a coherent interpretation. Its coherence is spurious since it wholly derives from contaminated evidence.[16]

(2) *The unity of mind argument*: The interpretation of dreams runs to a great extent parallel to diagnosing the origin of neurotic disorders. In the latter case the results arrived at are corroborated by a successful resolving of the associated symptoms (cf. Freud 1900, 528).

This argument can be unpacked as follows. First, the observable characteristics of dreams and psychopathological disorders have a lot in common. They likewise display apparent breaches, superficial allusions, unfounded transitions and so forth (cf. Freud 1940, 172). Second, similar effects ought to be traced back to similar causes – as Newton's second rule of reasoning requires (cf. Newton 1726, 398). Third, if the cause is empirically ascertained in one type of these apparently similar cases, we are justified in attributing the same cause to the analogous type.

The unity argument makes the adequacy of dream interpretations wholly dependent on the correctness of Freud's psychopathology. The latter's empirical success, however, has turned out to be fairly limited (cf. Grünbaum 1984, 231). And then the trouble lies with the third item: The cause has not been empirically ascertained in one of the related cases. With the demise of Freudian psychopathology, the Freudian theory of dreams topples as well.

CHAPTER IV

REICHENBACH LOOPS IN OPERATION: THE CONVENTIONALITY OF PHYSICAL GEOMETRY

In the next two chapters I work out my chief example of a complete theory, namely, physical geometry. The bulk of the present chapter is constituted by the discussion of the *conventionality of physical geometry*. Unlike the issues of testability and confirmation, addressed in chapter two, the conventionality problem has not been at the focus of attention of the philosophy of science community. Quite the opposite. A lot of people regard it as a dead issue. Still, it actually crops up time and again and seems to linger on in the wings. It appears to represent a robust third-rank issue.

The reason to turn to this problem in the present context is that some versions of the conventionality claim are founded on an untestability contention which can be reconstructed as arising from the recognition of Reichenbach loops. The relevant quantities are introduced on the basis of hypotheses that are potentially complete on the hypothesis level with respect to these quantities. The conventionality problem can thus be construed as a manifestation of the problem of testing complete hypotheses, so that an exploration of the former sheds additional light on the latter.

In the course of this exploration we will encounter old acquaintances, namely, non-instantiative testing and external fixing, the familiar antidotes against Reichenbach loop circularities; and we will come to realize to our regret that the impact of these strategies is fairly limited under these circumstances. The strategies are of no avail for solving the conventionality problem proper, the reason being that neither an agreement about acceptable constraints nor about the nature of the relevant pure states can be reached. Yet it will turn out that there is a Reichenbach loop related to this problem that can actually be tackled using these strategies. Either way we can learn a lot about the options for and the limits of testing potentially complete hypotheses by looking into the conventionality problem.

So, what will I do in this chapter is to exploit arguments adduced in favor of the conventionality claim for the purpose of this study. These arguments help to clarify the strengths and faults of the strategies developed in chapter two for defusing Reichenbach loop circularities. They expose the conditions on which the applicability of these strategies rests. It is fair, accordingly, that I try to make up for what I have taken from conventionalism; and that's what I will do in chapter VI. There I will address the question whether space-time geometry can be considered conventional in an interesting sense. And the answer to this question will be affirmative. But first I have to sketch the conventionality problem in its classic form. So allow me to begin by rattling old skeletons. But in order to prepare the stage for this performance, first a preliminary.

1. PRELIMINARY: NON-EUCLIDEAN GEOMETRY

§ 1. Not until it was finally proved that Euclid's fifth postulate, the notorious "parallel postulate", was in fact independent of the remaining corpus of Euclidean geometry, could the problem of physical geometry arise. Since Euclidean geometry constituted the only known geometrical structure before, it was a matter of course that the geometry of physical space is Euclidean in kind. As a result of the independence proof, as given by John Bolyai and Nicolai Lobachevsky in the 1830's, the parallel postulate could be denied and supplanted by a different claim without thereby necessarily running into a contradiction. As it turned out, there exist several ways for replacing consistently the parallel postulate, so that the unique Euclidean geometry gives birth to a variety of non-Euclidean geometries. Intuitively, these non-Euclidean geometries describe the geometric properties of curved surfaces rather than of flat planes.[1]

The first general account of the geometric relations on curved surfaces was developed by Carl F. Gauss. This account proceeded by embedding a curved two-dimensional surface in a three-dimensional Euclidean space and by describing the relations on the former with reference to a Cartesian coordinate system attached to the latter. Gauss succeeded in demonstrating that the embedding in Euclidean three-dimensional space can be dispensed with in principle. That is, the geometric relations on the surface can be determined by relying exclusively on quantities that are accessible on the surface itself.

The difficulty here is that due to the curvature of the surface a usual Cartesian coordinate system with its perpendicular straight axes cannot be applied to it. Gauss' solution was to introduce a coordinate system consisting of lines on the surface in question; and that means, he used curved lines as his coordinate axes. In particular, he introduced two such families of non-intersecting lines which covered the surface with a sort of bumped, crooked and bent grid structure. These lines are parametrized in a monotonic fashion. That is, if a given curve lies between two lines of this makeshift coordinate system, then the curve's relevant coordinate number lies between the parameter values associated with the coordinate lines. In this way every point on the surface can unambiguously be ascribed an ordered pair of coordinate values that circumscribes that point's location on the surface. On the other hand, this coordinate system is highly arbitrary, so that the coordinate values attached on its basis are unsuited for determining geometric quantities such as length. These quantities would vary with the coordinate system chosen.

In order to overcome this difficulty Gauss appealed to the Cartesian coordinate system in the embedding Euclidean space. In a Cartesian coordinate system the length or distance Δs can be derived from the differences in the respective coordinate values Δx_i using the three-dimensional variant of the Pythagorean theorem:

$$\Delta s^2 = \sum_{i=1}^{3} \Delta x_i^2$$

Now the trick is to express these Cartesian coordinate differences Δx_i as a function of the chosen internal coordinates u. In this way the Cartesian coordinate differences in the embedding space are translated into intervals within the curvilinear coordinate system attached to the surface. It is clear, to be sure, that this procedure is not applicable in a global fashion; the non-vanishing surface curvature vitiates the overall application of one comprehensive coordinate system with straight axes. Instead, the procedure has to be applied locally. That is, restricting oneself to tiny areas of the surface the distorting effects of surface curvature become small or, in infinitesimal regions, negligible. For these tiny or infinitesimal areas the internal coordinatization can be mapped into the coordinatization of the embedding Euclidean space. We have to use different maps for different regions, and in this sense we proceed locally. Gauss' chief discovery consisted in the proof that one single function (or rather three connected functions), whose values vary in general with the position, is sufficient for that purpose. This function is the *metric tensor* $g_{ik}(u_1, u_2)$. If this tensor is given, distances can be calculated on the basis of the surface-internal coordinate system (cf. Sklar 1974, 27–34; Mainzer 1980, 157–158).

§ 2. In 1854, Bernhard Riemann generalized the Gaussian theory of curved surfaces to n-dimensional manifolds. Such manifolds are characterized by the fact that each point in them can be uniquely specified by n real numbers. In particular, the manifold can be described by recourse to coordinate systems that respect neighborhood relations among points. That is, the parametrization of the coordinate lines employed must be smooth and monotonic (just as in the Gaussian two-dimensional model). A manifold thus allows for the distinction between neighboring points (or events in case of a space-time manifold) and distant points (or events), and that means, between local and non-local properties. This feature can be expressed by saying that a manifold is endowed with a topological structure.

The introduction of the concept of length or distance into such a manifold closely follows the Gaussian model. Quite analogous to the two-dimensional case we express infinitesimal lengths by processing coordinate differences (or differentials) as given in terms of some internal coordinate system u with the help of the metric tensor g_{ik}. One arrives at a formula that is identical, *mutatis mutandis*, with the corresponding Gaussian expression for surfaces:

$$(1) \qquad \mathrm{d}s^2 = \sum_{i,k=1}^{n} g_{ik}\, \mathrm{d}u^i\, \mathrm{d}u^k$$

This expression is usually simplified by introducing Einstein's summation convention. This convention stipulates that for all indices appearing twice, once up and once down, summation is to be understood. Accordingly, the summation sign can be deleted in Equation (1), so that it simply reads:

$$(1') \qquad \mathrm{d}s^2 = g_{ik}\, \mathrm{d}u^i\, \mathrm{d}u^k$$

The distance between two points P and Q is given by the integral over the path connecting them.

$$s(PQ) = \int_P^Q \mathrm{d}s$$

In contradistinction to Gauss, Riemann did not appeal to the notion of an embedding space of higher dimensionality and Euclidean structure. What he did instead was to attach to each point of the manifold a Euclidean tangent space. This presupposes that in small regions of the manifold Euclidean geometry holds. In that case the Pythagorean theorem (which underlies Equation (1)) is applicable to such regions. The metric tensor g_{ik} emerging from this procedure is of paramount importance for two reasons. First, it can be derived from geometric properties accessible within the manifold, and that means, without appeal to an embedding space. Second, it determines (almost) all geometric characteristics of the respective manifold.[2] In particular, it allows for the definition of a quantity that reflects the curvature of the corresponding manifold, namely, *Riemann's curvature tensor* – R^{i}_{jkl} in the four-dimensional or space-time case. This quantity specifies the curvature at each point of the manifold in each direction. If all its components are identically zero, the curvature vanishes and the respective space-time is flat.

§ 3. In 1917, Tullio Levi-Civita introduced the concept of a *geodesic* into Riemannian geometry. Geodesics are the straightest lines possible in a given space-time manifold. They are generated by displacing a curve's tangent vector as parallel as possible. For ease of illustration consider the two-dimensional case. We are presented with a spherical surface, and we wish to determine a geodesic on it. For that purpose we first attach the tangent vector at the point and into the direction in question. Now the problem is as follows. If we transport this tangent vector along a curve on the surface such that it retains its direction with respect to the embedding three-space, that is, if we perform an ordinary three-dimensional parallel transport, then the resulting vector will obviously no longer be tangent to the end-point of the path of transportation. It follows that on curved surfaces we cannot sensibly invoke the usual three-dimensional concept of parallel transport.

What we can do, however, is to introduce the notion of the "most parallel displacement possible." This means that a tangent vector is transported such that during the transport it remains within the tangent plane at each point on its path. Accordingly, the vector is slightly tilted with respect to the embedding space when moved from one point to its neighbor; this tilting signals the curvature of the manifold under consideration. But since there is no additional, active, as it were, change in direction involved, it is legitimate to consider this kind of transportation as a parallel displacement.[3]

A geodesic is then a curve whose tangent vectors are parallel in the just given sense. Geodesics are autoparallels. Technically speaking, parallel transport can be characterized by making use of the so-called *affine connection* Γ^i_{jk}. It is constitutive of a geodesic that the following equation holds:

$$\text{(2)} \qquad \frac{\mathrm{d}^2 x^i}{\mathrm{d}\lambda^2} + \Gamma^i_{jk} \frac{\mathrm{d}x^j}{\mathrm{d}\lambda} \frac{\mathrm{d}x^k}{\mathrm{d}\lambda} = 0$$

(λ denotes the curve parameter chosen, i.e., $x^a = x^a(\lambda)$). With the notion of parallel displacement – as formally expressed by the affine connection – the concept of direction is introduced into the manifold. On its basis it is determined whether or not two vectors point into the same direction. The criterion is whether one vector can be generated from the other by means of parallel transport (adjusting lengths if necessary). If space-time is flat, a coordinate system can be chosen such that all Γ^i_{jk}-components are identically zero. In the general case of curved space-time this can always be achieved locally. This possibility expresses the fact that curved manifolds can locally be approximated by flat tangent spaces. Moreover, the affine connection is definable using the metric tensor (along with its first derivatives),[4] but it can also be introduced independent of the metric – as was first noted by Hermann Weyl (cf. Weyl 1923a, 14). In addition to the property of maximum straightness, a geodesic exhibits the feature of extremal length. Geodesics are the shortest (or longest, as the case may be) connections between two points.

2. THE EARLY CONVENTIONALITY ARGUMENT: HELMHOLTZ, POINCARÉ, CARNAP

§ 1. Non-Euclidean geometry provides us with a collection of abstract mathematical structures. So the question emerges naturally whether one of these structures truthfully applies to reality and which one it is. That is, which of them correctly portrays the geometric relations prevailing in nature. As it turns out, this question is easier posed than answered, for we are faced with the problem of applying the mathematical quantities to experience. That is, we need a connection between the abstract object, the pure state, as it were, and physically tangible measurement outcomes.

The difficulty lurking here was first noted by Hermann von Helmholtz in the late 1860's. Helmholtz focused on the problem of length determination. In the first place, length is a concept of mathematical geometry, as defined by Equation (1) in IV.1.§2. In order for that concept to become part of physical geometry, we have to attach empirical indicators to it. The relevant measuring procedure is, as Helmholtz argued, the employment of *rigid bodies*. We measure lengths by using rods that retain their shape during the measuring process, so that we can place them next to one another in order to evaluate a given distance. Obviously, this procedure crucially depends on the possibility of identifying truly rigid bodies:

> All our geometrical measurements thus rely on the presupposition that the measuring instruments which we take to be solid [i.e., rigid bodies], are really bodies of invariable shape. At least it is presupposed that they are only subject to those alterations in shape which are known to us; e.g., the small deformations resulting from a change in temperature or from the different effect of gravity on rods placed in different positions (Helmholtz 1870, 23, translation mine).

That is to say, truly rigid rods indicate the sought-for length relations directly. The theoretically given quantity "length" manifests itself faithfully in the results of rigid-rod measurements. If, on the other hand, we are actually presented with deformed and thus non-rigid rods, then we have to correct the distorting influences present by recourse to physical laws.[5]

And now the catch occurs. How is it possible to test, we must ask, whether or not a given rod is truly rigid after the corrections we consider necessary have been performed. The assessment whether this rod really retains its shape ultimately has to rely on a comparison of its length to other rods supposed to be rigid. And this opens the avenue for a peculiar theoretical maneuver. We introduce a position-dependent deformation to which all rods are liable in equal degree. Because our assessment of rigidity essentially makes use of a comparison among rods, a like deformation of all rods can be assumed without running into empirical difficulties. To be sure, as Helmholtz points out, such a modification in the concept of rigidity requires a corresponding adjustment of mechanical laws. If, namely, the velocity of a particle that is not subject to any force is ascertained using the redefined rigid rods, this velocity comes out position-dependent and thus variable. The point is, however, that by a suitable change in the law of inertia accordance with the laws of physics can be regained (cf. Helmholtz 1870, 29).

Helmholtz, to be sure, was not a conventionalist. He maintained that geometric relations are to be based on facts of nature. Especially relevant is the fact of the "free mobility" of bodies, the fact, that is, that measuring rods can be moved through space without a change in their length (cf. Helmholtz 1870, 19–21; cf. also Helmholtz 1868, 37–41). On the other hand, it appears from Helmholtz's own investigations that this "fact" is of a somewhat shaky nature. Anyway, it was not Helmholtz but Henri Poincaré who put geometric conventionalism on the agenda of philosophy.

§ 2. Similar to Helmholtz, but independent of him, Poincaré focused on the problem of relating abstract geometric concepts to concrete experiential results. We never experience geometric quantities directly; rather, we are only presented with relations among bodies (cf. Poincaré 1902, 101). The point of Poincaré's approach is that these latter empirical relations may only provide a distorted picture of the underlying geometry. That is, Poincaré systematically considered distortions of geometric measurements and discussed their impact on the theoretical conclusions.

Imagine some creatures, as Poincaré's argument runs, who live in a world confined by spherical boundaries. Moreover, a temperature gradient is present in this world; in particular, the central regions are hotter than the peripheral ones. Another feature of this peculiar possible world is that the rate of thermal expansion is alike for all materials. Now suppose our creatures wish to determine the prevailing geometry. As a result of the variations in temperature, the length of the measuring rods will actually differ when used in the central regions as compared to the outward ones. And if our creatures do not consider and correct this length alteration and take their measuring results at face value, then they arrive at a geometry that is distinct from the one actually realized. That is, if they mistakenly assume that their measuring rods are rigid, then their outcomes misrepresent, say, the actually present Euclidean geometry as a non-Euclidean one.

The natural question at this juncture is: How can our misguided creatures come to realize that they are mistaken? And the sad answer is that they will never realize. For the empirically obtained length relations can *equally* be described by assuming a Euclidean geometry together with deformed rods or by assuming a non-Euclidean geometry along with rigid rods. This result remains unimpaired if straight lines or geodesics are taken into account. Geodesics are supposed to be physically represented by light rays. Then, as Poincaré argues, we can introduce an index of refraction that exhibits the same position dependence as the temperature gradient. In the case discussed by Poincaré, it follows that light rays no longer propagate along straight lines but rather along circular ones. Again, either we take this effect into account and obtain Euclidean geometry together with a supposed general perturbation of light propagation; or we interpret light rays as geodesics and arrive at a non-Euclidean geometry.[6]

Accordingly, Poincaré alters the geometric axioms and at the same time the empirical interpretation of the terms appearing in these axioms. The combined effect of these two alterations is a precise null result. Nothing of an empirically detectable nature is changed. What matters are not, of course, the fictitious inhabitants of bizarre possible worlds. It is to us that Poincaré's argument is really aimed. For in his view we are in exactly the same situation. That is to say, by relying on our usual notions of rigid rods and straight lines we could in principle come across indications that a non-Euclidean geometry is realized. We could, for instance, detect that the sum of the angles of a large, cosmic triangle formed by light rays does not add up to 180°. But we could easily explain away this indication by assuming that light rays do not follow straight lines but curved paths instead. For this reason, it is impossible in principle to ascertain empirically which geometry is actually realized in our universe. Consequently, this geometry is to be decreed by *convention*. It is Poincaré's prediction – which goes beyond the conventionality claim proper – that we will stick to Euclidean geometry come what may. For reasons of convenience and simplicity we will always make the conventional decision such that Euclid is retained forever (cf. Poincaré 1902, 95–97).

Poincaré did not intend to suggest that regarding physical geometry aspects of convenience ought to replace empirical criteria in the decision between alternative geometric variants. Rather, he considered these alternatives not as picturing different states of affairs but as describing the same state of affairs from different angles. The question as to which of these alternatives gives a true description of reality is comparable to the question if Cartesian coordinates are more true than polar coordinates (cf. Poincaré 1902, 76). Actually, there is no need to single out a coordinate system at the expense of an alternative one. And if we pick one, this choice is legitimately governed exclusively by pragmatic reasons.

§ 3. Poincaré's conventionality claim was taken up and further refined by Rudolf Carnap in his doctoral thesis. Carnap drew on a distinction worked out shortly before by Weyl, namely, the distinction between projective or affine geometry and metric geometry. I already alluded briefly to the fact that geometry need not be centered on the concept of length. Rather, one may operate on a narrower conceptual basis (cf. IV.1.§3). The central concept of affine geometry is displacement. Displacement can be represented by a vector since a vector connects two points, namely, its tail and its tip. In affine geometry vectors are supposed to form a so-called linear vector space. In particular, the concepts of the addition of two vectors and of the multiplication of a vector by a scalar are specified in that framework.

On that basis the concept of parallel displacement can be defined that in turn leads to the concept of a geodesic (cf. IV.1.§3). An affine space is identified by the corresponding family of geodesics which can be characterized by an affine connection Γ^i_{jk}. Two manifolds with the same class of geodesics have the same affine connection. The length of a vector is not, in general, defined in affine spaces. However, there is one special case in which length ratios can be obtained, namely, in case of parallel vectors. The multiplication of a given vector by a scalar is to be interpreted as an expansion or contraction of this vector, and by combining this procedure with a parallel displacement of the vector it is possible to compare its length with another one parallel to the first. The length ratio is given by the corresponding scalar factor.

The introduction of the general concept of length is based on the concept of scalar product. A vector's squared length is the scalar product of this vector with itself. If this latter conception is adduced to the postulates characterizing affine spaces, we obtain metric spaces. Metric spaces are characterized by a metric tensor g_{ik}, and on its basis arbitrary lengths are fixed (up to a choice of units). In general, there are several options for introducing a metric tensor into a given affine space. That is, a given class of geodesics is compatible with various distinct metrics. On the other hand, a given metric unambiguously determines a class of geodesics. The metric structure fixes the affine structure but not the other way around.[7]

Carnap focuses on the problem of ascertaining empirically which lines are straight or, in the more general case, geodesic. In view of the just-described hierarchy of geometric structures, he discusses relevant procedures on either level. The first option is to directly single out natural objects or processes that are apt to represent straight lines. The second option is to rely on bodies which are to be considered rigid and which can thus be employed for measuring lengths. Since the metric fixes the affine connection, straight lines are given once rigid rods are available. The shortest connection between two points is the straightest one possible.

The problem with the second option though is that the rigidity of rods is not a simple fact of nature – as already stressed by Poincaré. Whether or not a rod retains its length can ultimately only be determined by comparing it to some other rod whose rigidity has to be ascertained by way of another comparison to yet another rod and so on. The attempt to ascertain rigidity empirically thus gets stuck in an infinite regress. Consequently, there is room for adopting different standards of rigidity which respectively lead to conceptually different but empirically equivalent accounts of physical geometry (cf. Carnap 1922, 33–37).

Carnap gives a concrete elaboration of this abstract possibility by working out the consequences of the postulate that the earth's surface is a flat Euclidean plane. In order to obtain measuring results in agreement with this postulate we have to assume that the lengths of all rods are dependent upon their respective positions or upon their height above the surface. Since the measured raw data do not, of course, reflect this variability, we have to apply a corrective procedure in order to sift the true distances out of these raw data. Carnap explicitly specifies such a procedure, i.e., a law for the variation of length with position, in whose light the earth's surface appears flat. In addition, the earth appears to be located in a universe of finite extension; very finite indeed, since the largest possible distance of a star comes out as roughly 10,000 km.

It is clear that this remetrization implies some changes in the behavior of objects other than rods. For instance, in order to maintain empirical equivalence it is to be required that light rays move along circles. This requirement can be fulfilled by applying Poincaré's recipe: We assign to empty space a position-dependent index of refraction. With this boundary condition the usual wave optics can be retained. Moreover, if we stick to the received mechanical principles the remetrization implies that familiar theorems, such as the conservation of mechanical energy, no longer hold. It is possible, however, to repair this defect by way of a redefinition of mechanical quantities. If the equation of motion is appropriately altered and the definition of kinetic energy properly adjusted, the conservation of mechanical energy can be derived – just as in the usual version. In addition, Hamilton's principle remains valid under these circumstances (cf. Carnap 1922, 47–54).

The same holds analogously with respect to the first option, namely, the direct introduction of a preferred set of geodesic lines. Suppose we wish to determine whether a given rod is actually straight. This can be done by looking along the rod, i.e., by comparing its edge with the path of a light ray. But how do we know that such rays are legitimately considered straight? In fact, as Carnap argues, it is impossible to know. Restricting ourselves to what is given in experience, we are prevented in principle from telling apart what is straight from what is curved. The upshot is, then, that we have to decree by convention what physical objects or processes are to pass as rigid or straight (cf. Carnap 1922, 33, 37).

Like Poincaré, Carnap specifies criteria that are supposed to direct the conventional decision necessary at this juncture. And just like Poincaré again, he appeals to convenience and simplicity; more precisely, to Mach's principle of the economy of thought. But the way he brings to bear this general approach, as well as the conclusions he arrives at, differ considerably from Poincaré's. The point is, as Carnap argues, to apply the criterion of simplicity or economy not to geometry in isolation. This is what Poincaré had done in order to distinguish Euclidean geometry. Rather, this criterion is to be applied to the whole system of physical laws that emerges from the choice of a particular geometry. The standard of rigidity is to be picked so that the resulting overall structure of physical laws takes the simplest form possible.

Moreover, a rule can be formulated that appears as a promising candidate for capturing the spirit of this approach. Namely, if it turns out that all bodies display a numerically identical behavior in some respect, the principle of economy of thought advises us to explain away this numerical coincidence by an appropriate redefinition of the standard of reference. That is, the assumption of a deformation to which all rods are subject in equal measure should be rejected in favor of a correspondingly adjusted standard of rigidity (cf. Carnap 1922, 36, 55–59).

This means, if we hit upon empirical indications that a non-Euclidean geometry is physically realized, we should not override these indications by introducing general deformations of the lengths of rods and the paths of light rays. On the contrary, the principle of overall simplicity might be heeded best by giving up Euclidean geometry. The point of this recommendation is that since Poincaré's time the relevant physical situation had changed drastically. In the mean time Einstein's theory of General Relativity had been formulated and had gained acceptance. And as a matter of fact, Einstein had been in a situation comparable in some respect to Poincaré's scenario. However, Einstein had opted differently than Poincaré had foreseen and adopted a non-Euclidean geometry. Carnap's point in his recommendation is to capture and thus justify Einstein's approach. In order to see what is at issue here, a brief digression on the basics of General Relativity Theory is in order.

3. FUNDAMENTALS OF GENERAL RELATIVITY THEORY

In order to prepare the physical ground for the philosophical discussion that is to follow it is appropriate to outline the fundamental principles of General Relativity Theory. I focus on the general ideas underlying the theory and omit most of its technical subtleties. The presentation essentially follows Einstein's own style of developing and interpreting the theory.

§ 1. Einstein introduced General Relativity on the basis of three principles, namely, (1) *Mach's principle*, (2) the *principle of equivalence*, and (3) the *principle of general covariance*. Mach's principle is intended to express the requirement that there is no absolute space. On the Newtonian view, absolute space constitutes a preferred frame of reference ("absolute rest") and thus provides a basis for distinguishing among different forms of motion; in particular, it allows for distinguishing between uniform rectilinear motion, i.e., inertial motion, and accelerated motion or rotation. A body is moved in a certain fashion with respect to the preferred frame of absolute rest. The inertial forces accompanying non-inertial motion are interpreted as resulting from the motion with respect to absolute space. Accordingly, there exist tangible effects of space on moving bodies. On the other hand, these moving bodies are unable to influence in turn the spatial structure. Einstein felt that such a one-sided causal relation was contrary to scientific reasoning. Absolute space should be supplanted by the assumption that the inertial properties of moving bodies are determined by a physical field analogous to the electromagnetic one. This means that the metric and geodesic structure of space-time should be determined completely by the relative positions and motions of material bodies; put conversely, space-time should not possess an intrinsic structure independent of the location of bodies (or fields). This is the assumption Einstein called Mach's principle.[8]

The second axiom is the principle of equivalence. Experience shows that inertial mass (signaling the acceleration value a body receives from a force of given intensity) and gravitational mass (indicating the intensity of the gravitational field generated by a body) always coincide numerically. This numerical agreement becomes manifest in the fact that all bodies are accelerated exactly alike in a gravitational field regardless of differences in their mass values or their chemical composition. The numerical agreement of inertial and gravitational mass is already part of Newtonian mechanics, but it is a purely accidental feature in that theory and thus remains unexplained by it. An explanation of the numerical coincidence has to interpret it as the manifestation of a deep-rooted identity of inertial and gravitational mass. One has to assume that one and the same property expresses itself as inertia or as gravity depending on the circumstances (cf. Einstein 1917, 53–54; Einstein 1922, 56–57). The assumption of the equality or identity of inertial and gravitational mass is called the *weak principle of equivalence.*

If the weak principle of equivalence holds the inertial and gravitational forces acting on a freely falling particle exactly compensate each other. Consequently, no effect of the gravitational field on the particle can be detected empirically. This means that it is impossible to decide on the basis of mechanical experiments whether an observer is freely falling in a uniform gravitational field (in which the field's direction and intensity are spatiotemporally constant) or whether she moves inertially through a field free region. In both cases, namely, the motions of all bodies (if not subject to additional forces) are independent of their mass and their internal constitution and follow parallel paths. If one sets aside for the moment the question as to the sources of the gravitational field, an inertial frame of reference is entirely equivalent mechanically to a frame falling freely in a uniform gravitational field. In a spaceship moving uniformly through empty space all motions proceed just as in a freely falling elevator near the earth's surface.

The assumption that this equivalence is not confined to mechanical processes but rather extends to all physical processes is called *strong principle of equivalence.* According to the strong principle, the two types of reference frames are equivalent with regard to all physical processes, and in particular, with regard to optical phenomena such as the paths of light rays. That is, the propagation of light, too, does not provide a basis for distinguishing between a uniformly accelerated frame and an inertial frame.

Among Einstein's chief motives for formulating and emphasizing the principle of equivalence was to remove the Newtonian concept of inertial frame from physics. The introduction of privileged inertial frames of reference, as in classical mechanics, appeared illicit to him since these privileged frames could not be singled out empirically in an unobjectionable way. The problem with inertial frames is that, on the face of it, an operational circularity is involved in their selection. The law of inertia states that the acceleration of a body moving freely, i.e., uninfluenced by any force, vanishes. But in order for that statement to become empirically meaningful, its application has to be restricted to particular frames of reference. For the acceleration of arbitrarily moved bodies can be made to vanish by using a frame of reference moved along with the body in question. On the other hand, the physical realization of the right sort of frames, i.e., rectilinearly and uniformly moved frames, demands recourse to bodies not liable to external forces; bodies, that is, moved inertially. In response to this apparent circularity, the law of inertia is often given the form of an existence claim: Frames of reference can be physically realized in which bodies uninfluenced by any force move without acceleration.

This response, however, was frequently considered less than convincing, and so Einstein's remedy to this difficulty was to point out that in light of the principle of equivalence the whole distinction between inertial and non-inertial frames loses importance.[9] For the boundary between inertial and non-inertial frames can be shifted by the introduction of gravitational fields. Instead of referring to inertial frames in a field-free space, one might just as well speak of non-inertial (namely, freely falling) frames in a gravitational field. Accordingly, what is to qualify as an inertial frame can only be determined after the structure of the gravitational field has been established. This implies that inertial frames do not denote an intrinsically distinguished state of motion; inertial frames are not privileged by nature.[10] Since the occurrence of inertial forces is crucially dependent on the nature of the frame of reference chosen and since the nature of that frame is conditional on the gravitational field, it follows that the occurrence of inertial forces is connected to gravity.

This consideration leads to the third Einsteinian axiom, namely, the principle of general covariance. Einstein's concern was to avoid the introduction of any privileged class of frames of reference. Consequently, the theory had to be formulated so that its equations are applicable as they stand in arbitrary coordinate systems. The class of admissible coordinate systems may at most be constrained by the condition that spatiotemporally nearby points be ascribed neighboring coordinate values. Einstein believed that this principle of general covariance could be considered a concrete implementation of the general principle of relativity which states that no preferred kinds of motion exist (cf. Einstein 1922, 61).

As a matter of fact, however, there is no such connection between general covariance and the relativity of motion, as Erich Kretschmann pointed out in 1917. Kretschmann showed that every physical theory can be given a generally covariant form. General covariance is thus a mathematical rather than physical property of a theory; it refers to the style of formulation rather than to physical content. In particular, both Special Relativity and Newtonian mechanics can be cast in a generally covariant fashion. Thus, general covariance does not entail the physical equivalence of frames of reference.[11]

§ 2. After this exposition of the three principles Einstein invoked, let us take a closer look at their use and mode of application. The principle of equivalence has been applied to non-uniform gravitational fields as well. Its extension to arbitrary fields, however, requires a corresponding restriction in the spatiotemporal regions in which it is supposed to hold. In non-uniform fields it can only be applied *locally*, i.e., to events sufficiently close to one another. Think of the earth's spherically symmetric gravitational field and take two particles released with a significant horizontal separation. Since both trajectories are directed at the earth's center the two particles will approach one another. And this means, they do not pursue parallel paths, as they would if they were moved inertially. So, the changing intensity of the field cannot be accommodated by the introduction of one comprehensive (or global) freely falling frame of reference. Accordingly, not all gravitational fields can be transformed away through the transition to an accelerated frame. Rather, the global equivalence between inertial and freely falling frames is confined to uniform fields (cf. Einstein 1917, 57). Still, if consideration is confined to a small neighborhood of a given event, this disturbing effect becomes negligible. In its local version the strong principle of equivalence states that at every space-time point in arbitrary gravitational fields a frame of reference can be chosen such that in a sufficiently small neighborhood around the point no effects of gravitation occur. A freely falling observer restricted to processes occurring nearby feels nothing of a gravitational field around him.[12]

With the help of the principle of equivalence, some effects of a gravitational field on the propagation of light can easily be clarified. Consider a beam of light that is emitted from the wall of a box accelerated upward. The beam passes through the box perpendicularly to the direction of motion with the result that it follows a curved path when registered by an observer at rest with the box. Now the strong principle of equivalence implies that the very same path is also followed if the box is at rest in a gravitational field. The conclusion is that light is deflected by such a field (cf. Einstein 1917, 60; Sklar 1974, 69).

The general technique deployed here is as follows. In order to find out what happens in a gravitational field, go over, first, into the locally equivalent accelerated frame of reference, i.e., into the so-called *local inertial frame* in which no gravitational field is detectable. This local inertial frame is physically realized by a freely falling body at the space-time location in question. Second, analyze the relevant processes in the framework of Special Relativity. Subsequently, third, translate the results gained in the second step back into the original frame of reference (cf. Einstein 1916, 118–120; Sklar 1974, 69–70).

The same recipe is applied in the derivation of gravitational red shift and time dilation. Consider again a box in upward acceleration. From its bottom a light beam is sent off in the direction of motion. When the light beam reaches the top of the box, the box moves faster (due to its acceleration) than it did when the light was emitted. Since the velocity of the light source when it emits the beam is smaller than the velocity of the detector when the beam arrives, a Doppler red shift occurs. Transferring this result to the gravitational-field case leads to the result that the frequency of a light beam running upward in a gravitational field is decreased as well. Moreover, if one takes the time between the arrival of two successive wave crests as ticks of one's clock, a decrease in frequency signals a slower rate of this clock. So if an observer located at the top compares the time between the incoming crests with the temporal intervals as furnished by a clock stationed next to her, she will conclude that the clock at the bottom runs at a slower rate. This means that a clock at rest within a stationary gravitational field is slowed down with respect to a clock outside of it or with respect to a freely falling clock.

§ 3. From clocks to rods. We can apply the same general procedure in order to elucidate the behavior of rods in gravitational fields. Consider a rotating disk in a gravitation-free region. We wish to ascertain empirically the geometric relations on the disk, and we employ measuring rods for that purpose. Now the following peculiarity occurs. Rods placed in the direction of the circular motion, i.e., tangentially to the rotating disk, are subject to a Lorentz-contraction, whereas rods placed perpendicular to that direction, i.e., oriented radially, remain uninfluenced by this effect.

Now we have the choice between two alternative options for describing this scenario. Either we consider the measuring rods placed tangentially as contracted and accordingly correct the shrinking by applying Lorentz's transformation law to the raw data. In that case we arrive at the familiar Euclidean geometry. Or we assume that all rods on the disk retain their length regardless of their orientation. That is, we consider the uncorrected measuring results as truthful indications of the prevailing geometric relations. On that approach the ratio between the circumference and the radius of a concentric circle on the disk will obviously diverge from its Euclidean value 2π. This means that the geometry on the disk comes out as non-Euclidean. The conclusion is that on the basis of the constancy assumption for rods the geometry in rotating frames of reference, or to generalize somewhat, in non-inertial frames of reference, is non-Euclidean in character.

Again, the final step is to transfer this result to an inertial frame within a gravitational field. That is, we assume that the disk is non-rotating and that a suitably shaped gravitational field is present on it. The conclusion is, then, that a gravitational field manifests itself in a change of physical geometry. The presence of such a field finds its expression in geometric relations of a non-Euclidean type in the region under consideration. Gravitational fields are nothing but a particular structure of physical geometry.[13]

It has to be borne in mind that this result is only obtained if we deliberately renounce the correction of measurement results for gravitation-induced deformations. That is, if we refrain from considering gravitational influences on measuring rods as distortions in need of rectification. This approach derives its plausibility from the fact that the alternative would require a correction procedure that is identical for all rods. The deforming influence of the gravitational field assumed in this alternative scenario is independent of the internal constitution of the rods employed and is thus equal for rods of any material. It is tempting (and in conformity with Carnap's rule as given in IV.2.§3) to explain away such a general deformation by the assumption that gravity alters the space-time structure which is then truthfully indicated by rod measurements.

It is clear that such an attempt to "geometrize" the effects of a gravitational field cannot be restricted to a reinterpretation of the behavior of rigid rods; rather, it must be extended to the motion of particles. If gravity is to be incorporated in the space-time structure we arrive at the following version of the law of inertia: Particles in a gravitational field, and with no other forces present, move along the straightest line possible in the respective space-time. They move along geodesics. It is important to realize that these geodesics are not to be understood as the straightest lines in space but in space-time. The elliptic orbit of a particle in a spherically symmetric gravitational field certainly does not represent the straightest spatial line. What is at issue, however, are not three-geodesics but four-geodesics, which indicate not only a spatial direction but a velocity as well. A spatiotemporal or four-geodesic characterizes a motion in a certain spatial direction with a certain velocity. It is a consequence of the weak principle of equivalence that the inertial motion thus described is unique in the following sense: At each space-time location in each four-direction one and only one inertial four-trajectory exists. It is by virtue of this uniqueness that such trajectories can be construed as geodesics. For otherwise the geodesic structure would depend on some particle parameter which obviously makes no sense.

§ 4. Let me couch the geometric approach to gravitation a trifle more technically. Inertial four-trajectories are characterized by the condition that the acceleration of a particle moving along them vanishes. That is:

$$\text{(3)} \qquad \frac{\mathrm{d}^2 x^\alpha}{\mathrm{d}\tau^2} = 0$$

(with proper time $d\tau$). In this form, however, the equation is only applicable within inertial frames x^α (see § 1), and so we have to give it a generally covariant shape. This is done by introducing an arbitrary coordinate system x^i and by specifying the coordinate transformations that relate x^α and x^i. In this way one arrives at the following generally covariant formulation of Equation (3):

$$(4) \qquad \frac{\mathrm{d}^2 x^i}{\mathrm{d}\tau^2} + \Gamma^i_{jk} \frac{\mathrm{d}x^j}{\mathrm{d}\tau} \frac{\mathrm{d}x^k}{\mathrm{d}\tau} = 0$$

Equation (4) coincides with Equation (2) (cf. III.1.§3) if the curve parameter λ is substituted by proper time τ, i.e., if the trajectory is parametrized using proper time. So we recognize in Equation (4) the generally covariant description of a geodesic. Equation (4) thus describes without recourse to any distinguished frame of reference the motion of a particle whose acceleration vanishes in the local inertial frame.

In order to arrive at a more detailed physical interpretation of Equation (4), let's take a closer look at it. A generally covariant description is given by using tensors. Tensorial quantities are characterized by a certain behavior under coordinate transformations. In particular, tensor equations remain intact if a different coordinate system is adopted. This implies that if a tensor vanishes in one coordinate system, it vanishes in all coordinate systems. The left-hand side of Equation (4), in virtue of being a tensorial quantity, vanishes in all coordinate systems. Things are different, however, as regards the two terms on the left-hand side separately. None of these terms in isolation, but only their combination, is tensorial in character. In local inertial frames Equation (4) goes over into Equation (3). From this it appears that in these latter frames the affine connection Γ^i_{jk} vanishes. This property is constitutive of local inertial frames. One recognizes, then, that on the whole such frames are characterized by the following bundle of traits: In local inertial frames no gravitational field is detectable; consequently, the metric in them coincides with the Minkowski-metric of Special Relativity; local inertial frames are physically realized by freely falling particles; they are mathematically expressed by the property that all components of the affine connection are identically zero.

In view of these features, the following reasoning appears plausible. Since the transition to the local inertial frame makes gravitational effects disappear and at the same time reduces the affine connection to zero, we may interpret the affine connection as a mathematical representation of the gravitational field. More specifically, we construe the second term of the left-hand side of Equation (4) as representing the gravitational acceleration or field strength, and its first term as inertial acceleration.[14] The missing tensorial nature of each of these terms considered in isolation can now be thought as signaling the important fact that gravity and inertia do not possess separate physical significance. The split between these two quantities is conditional on the frame of reference chosen. Accordingly, it is always possible to transform away the gravitational field (at least locally) by picking a particular frame (the local inertial frame), and this demonstrates that gravitation has no independent existence. Only the combination of inertia and gravity is a frame-invariant quantity and is thus to be considered physically significant (cf. Einstein 1922, 81–82).

§ 5. On the whole, then, General Relativity proceeds by geometrizing gravitation: It is built into the space-time structure. For this reason gravitation is attributed a special status in that theory; it is construed as being different in kind from all other interactions. Whereas electromagnetic and similar forces are interpreted as altering a rod's length and as deflecting a particle from a straight line, things are different with regard to gravity. Gravity embodies the structure of physical geometry, so that its influence on rods, clocks and particle trajectories is not viewed as distortion and is consequently not corrected. After all, it makes no sense to correct the influence of space-time on space-time measurements.

With respect to trajectories this means that freely falling bodies are considered force-free. Free fall is the natural state of motion in the sense that no forces are involved in it. Accordingly, freely falling bodies move along the straightest lines possible; they follow the geodesics of the corresponding space-time. These geodesics thus represent a generalized concept of inertial motion which takes into account that gravity and Newtonian inertia are inextricably interwoven. The geodesics constitute the affine structure of space-time and are given by Equation (4).[15] On this interpretation the law of inertia expresses this affine structure and has nothing to do with an existence claim for inertial frames (see § 1) (cf. Earman/Friedman 1973, 337–338; Ehlers 1973a, 75).

According to a well-known legend, it was the impact of an apple falling from a tree which led Newton to the idea that gravity was a universal and reciprocal attraction among bodies. In order to get the flavor of geometrization let's see what picture of the falling apple emerges in General Relativity. First, since the apple is freely falling its motion is force-free. The apple is not attracted by the earth; it simply follows the straightest trajectory available to it. Second, it is rather the tree that moves under the influence of forces (cf. Synge 1960, 132–133). More specifically, the trajectory of the tree is bent away from the geodesic (as represented by the motion of its former component) through the action of cohesion. It is cohesive forces that fix the tree's location on the earth's surface and thus prevent it from assuming its natural state of motion. So, the tree is kept from joining the apple's motion, and the apple is not driven away from the tree. Since the only force involved is cohesion, it is cohesion, and not gravity, that is the primary cause of the two bodies' growing separation.

Likewise, what we feel when sitting at a desk is not the force of gravity. It is again the force of cohesion that makes the parts of the chair (and of the earth below it) stick together and thus blocks our entering the natural and distinguished state of geodesic motion. These two examples bring to light that geometrization really makes a difference to our understanding of gravitation.

§ 6. One important aspect of General Relativity has been left untouched so far, namely, the non-Euclidean nature of physical geometry. Geometries of the non-Euclidean variety are distinguished by the fact that they allow for deviations from flatness. Within their framework curvature makes its appearance – in the case of physical geometry, curved space-time. In General Relativity space-time curvature is of chief importance. It determines the behavior of rods and the paths of particles and light rays. Conversely, space-time curvature is generated by matter, electromagnetic radiation, mechanical stress and the like. This action of matter on space-time structure is constitutive of the geometric approach to gravity; if gravity is supposed to be expressed by space-time structure, the presence of matter (and other sources of the gravitational field) must be able to alter that structure.

This action of matter on space-time is accommodated by Einstein's *field equations* of the gravitational field. These equations contain the so-called stress-energy tensor T^{ik}, which captures all non-gravitational sources of the gravitational field (namely, momentum and energy of all non-gravitational forms of energy). The Einstein equations relate this stress-energy tensor to the metric and curvature of space-time.[16] The field equations represent the core of General Relativity. They are in a sense analogous to Newton's law of gravitation (or rather to its local version, Poisson's equation); in fact, this law is obtained from the field equations in the weak-field slow-motion approximation. But the explanatory power of Einstein's equations exceeds by far that of their Newtonian ancestor. These equations entail the local conservation of energy and momentum. Moreover, in contrast to their Newtonian analog, they need not be supplemented with an independent equation of motion. Rather, it can be derived from them that test particles (i.e., small, neutral and non-rotating particles) follow a geodesic path in a gravitational field; it can be derived, that is, that their motion obeys Equation (4). Moreover, it is also a consequence of the field equations that light rays propagate along a distinguished subset of geodesics ("null geodesics"). Accordingly, the behavior of rods, clocks, particle trajectories and light rays, which was explained above on the basis of the principle of equivalence and Special Relativity, can also be derived from the field equations directly. It will be seen that this important feature contributes to the high degree of Einstein–Feigl completeness that General Relativity enjoys (cf. V.3.§3).

In contradistinction to the equations for the electromagnetic field, the gravitational-field equations are non-linear. This non-linearity arises from the fact that the gravitational field carries energy and momentum and hence itself constitutes a source of the gravitational field. The electromagnetic field, by contrast, is not electrically charged. In most practical applications, however, the effect of the gravitational field on itself is very small and can thus be neglected. In these cases the field equations can be linearized and then assume a shape akin to the equations of the electromagnetic field (cf. VI.2.§2).

§ 7. It is important to realize what precisely are the effects of curvature on particle trajectories and measuring rods. In General Relativity, curvature is associated with changes in the gravitational field strength. This means, conversely, that in a static and uniform gravitational field curvature vanishes. The most important indication of space-time curvature is *geodesic deviation.* This quantity expresses the change in separation between neighboring geodesics. Couched in physically more tangible terms it expresses the relative acceleration of two freely falling particles released at neighboring space-time points. The magnitude of geodesic deviation is related to the prevailing Riemann curvature and the initial separation of the geodesics. The larger one of these quantities, the larger the resulting effect. Geodesic deviation thus provides a locally accessible measure for the deviation from flatness (cf. Ehlers 1973b, 10;

Misner/Thorne/Wheeler 1973, 34, 265–270; Stephani 1980, 8–10).

Intuitively, geodesic deviation comes about as follows. In a gravitational field of varying intensity the two neighboring particles experience a different action of the field. Consider for instance the stationary and spherically symmetric (and thus non-uniform) field around an astronomical body (such as the earth, the sun or a black hole). Within that field the gravitational attraction varies with the radial distance from the center, so that two bodies at different distances are attracted with forces of different strength. The difference between these two forces induces a relative acceleration. Similarly, two nearby bodies moving jointly along the radially oriented field lines of such a field are likewise subject to a relative acceleration. In the most basic case of two neighboring freely falling particles, this acceleration is precisely the geodesic deviation. This effect is also known under the more familiar name of *tidal forces*. If tidal forces act on a spherical body they tend to deform it into an egg-shaped one. That is, they expand it in one direction and contract it in the perpendicular directions. In particular, they tend to elongate measuring rods placed radially in such a spherically symmetric field. The cause of this elongation is that the field exerts forces of different strength on the two ends of the rods.[17]

As already indicated, curvature is expressed mathematically by the Riemann curvature tensor and is thus a frame-independent quantity. This implies that curvature cannot be made disappear by switching to a different coordinate system. In particular, it does not vanish automatically in the local inertial frame (cf. Landau/Lifshitz 1976, 312). In full generality, curvature effects can only be made to vanish at a single point (but not in its small neighborhood) or along a trajectory (but not in the small region surrounding it) (cf. Friedman 1983, 200–202; Ray 1987, 67–73). The local vanishing of curvature depends on the prevailing physical circumstances and not on the adoption of a particular frame.

But now we are faced with a difficulty. As explained in § 5, General Relativity is characterized by the geometrization of gravitation, and this geometrization entails that the influence of gravity on the length measuring rods is left uncorrected. But what are we to make of curvature effects such as tidal forces against this background? After all, tidal forces definitely tend to contract or expand measuring rods; they may even tear them in pieces. Rods subject to such forces no longer qualify as rigid. Even in General Relativity the influence of tidal forces is regarded as a distortion and is consequently corrected. In other words, there are effects of gravitation that remain accessible within the local inertial frame. And this fact appears to militate against the local version of the strong principle of equivalence (see § 2); it casts doubt on the application of the principle to non-uniform fields.

It is true that most curvature effects become negligibly small in the local inertial frame. Since the magnitude of geodesic deviation (and consequently of tidal forces) is dependent on the separation of the particles in question (and consequently on the extension of the rod), this magnitude reduces to zero as the separation (or the rod's extension) decreases. That is, although curvature does not vanish in the local inertial frame it no longer leads to empirically detectable deformations. Midget rods are not significantly affected by tidal forces.

But this response won't do under all circumstances. There are effects of curvature which do not even approximately vanish in the local inertial frame. It has been pointed out by Hans Ohanian that although the absolute magnitudes of induced tidal deformations diminish with the shrinking extension of a three-dimensional configuration, the ratio of these deformations to the size of the configuration remains constant. In the case of a deformable liquid drop, for instance, one arrives at the result that the ratio of the induced tidal deformation to the drop's radius is completely unaffected by the decreasing size of the drop. And this means that the drop's shape is invariant under scale alterations in curved space-time (cf. Ohanian 1977, 906). From this and similar peculiarities Ohanian concludes

> that in some systems the effects of curvature never do disappear, not even in the limiting case of systems of infinitesimal size (Ohanian 1977, 903–904).

That is, curvature does not cease to have physically tangible effects in arbitrarily small regions; in particular, it remains empirically accessible in principle in the local inertial frame. In arbitrary gravitational fields the equivalence between freely falling frames and the inertial frames of Special Relativity only holds for point events. Contrary to what is often thought, freely falling Liliputians cannot forget about gravitation altogether.

§ 8. General Relativity suggests that we deny separate existence to the gravitational field and to incorporate its effects into space-time structure. But precisely what aspect of space-time structure is supposed to express these effects? As already mentioned, Einstein considered the affine connection, and that is, the spatiotemporal changes in the metric, to be the relevant quantity (cf. Einstein 1916, 120). This construal grows out of Einstein's conviction that at bottom inertia and gravity are identical. The frame-dependence of the affine connection shows that the distinction between inertial and gravitational effects is likewise frame-dependent and thus spurious and without objective significance (see § 4).

The problem with this construal lies in the fact that it can no longer unrestrictedly relate the gravitational field to its sources. After all, to the alleged equivalence between a uniformly accelerated frame of reference and an inertial frame in a uniform gravitational field one could object that, as the case may be, no sources of such a field are detectable and that, consequently, the first interpretation is objectively preferable to the second.[18] A uniformly accelerated box in a region free of masses and radiation should sensibly be construed as a uniformly accelerated box since no possible causes of a gravitational field are available.

Einstein tried to circumvent this objection by interpreting the gravitational field as an independent entity, i.e., as an entity that cannot be reduced to the role of a mere mediator of interactions among material bodies. General Relativity is essentially and irreducibly a field theory (cf. Einstein 1922, 140). This is to say, the fact that the gravitational field cannot in every case be traced back to material sources shows that the field has a sort of independent existence detached from material bodies.

In contrast to Einstein's view, John L. Synge has identified the gravitational field not with the affine connection but with the Riemann curvature tensor instead. "The Riemann tensor [...] *is* the gravitational field" (Synge 1960, VIII). The chief advantage of this interpretation is, as Synge argued, that curvature is frame-independent.

> In Einstein's theory, either there is a gravitational field or there is none, according as the Riemann tensor does not or does vanish. This is an absolute property; it has nothing to do with any observer's world-line (Synge 1960, IX; see also *ibid.* 109).

In a globally flat space-time there are no circumstances under which a gravitational field exists. The introduction of an accelerated frame of reference simply amounts to the introduction of an accelerated frame of reference and has nothing to do with the generation of a gravitational field. Accordingly, the only truly gravitational effects are tidal forces since they arise from space-time curvature (see § 7).

It is clear that on this approach the unity of inertia and gravity, which is central to Einstein's interpretation, dissolves. Inertial effects can be distinguished unambiguously from gravitational effects in Synge's sense. Synge's approach is supported by the fact that even in local inertial frames tidal forces are empirically accessible – with the result that the comprehensive empirical equivalence between inertial and gravitational effects does not hold anyway (see § 7). Moreover, curvature (i.e., gravitation in Synge's sense) is unambiguously determined by the sources of the gravitational field via Einstein's equations. Causality is thus preserved.[19] Accordingly, Synge's interpretation is preferable to Einstein's.

It becomes clear now that all three principles on which the development of General Relativity was based (see § 1) are somewhat shaky. Mach's principle, the principle of equivalence and the principle of general covariance have either not stood up to critical scrutiny or have failed to live up to the expectations initially attached to them. As the French say, *les fruits n'ont pas tenu les promesses des fleurs*. On the other hand, it turned out that the theory had virtues unanticipated at its inception. Chief among these is the discovery that Einstein's field equations actually imply the appropriate equation of motion. It came out as an unexpected present that the geodesic nature of free fall and light propagation need not be postulated separately.

Another gain in parsimony, this time an expected one, is due to the fact that General Relativity, in contrast to all alternative theories of gravitation, can dispose of a fixed background metric altogether. Whereas all of these alternative approaches have to employ two relevant quantities, namely, the gravitational field and the space-time container, General Relativity needs only one. The metric tensor suffices for a comprehensive description of space-time structure and gravitation alike. So, Ockham would have been pleased.

4. REICHENBACH'S CONVENTIONALITY ARGUMENT AND ITS IMPACT ON THE COMPLETENESS OF PHYSICAL GEOMETRY

After this excursion through the standard account of physical geometry I will now resume the discussion of the philosophical thesis that the structure of physical geometry is conventional in important respects. It is clear that this thesis is to be viewed and assessed against the background of the relevant physical theory, and this is what justified, and even required, the foregoing detour. The conventionality thesis was given its canonical form by Reichenbach. So, what I will do in this section is, first, to present his views on that matter, second, to outline an updated version of the thesis, and third, to relate the conventionality thesis to the main issue of this book, namely, Einstein–Feigl completeness and its problems. It will be seen that Reichenbach's formulation of the conventionality thesis heavily draws on the earlier work of Helmholtz, Poincaré and Carnap (as sketched in IV.2); but the conceptions imported from this tradition are considerably refined, elaborated and systematized.

§ 1. We start with the problem of ascertaining empirically the geometric relations realized in our surroundings. For this task we are in need of measuring rods of unchanged length. In particular, we are in need of rods that retain their length when transported. Only on that condition we can expect to obtain data that truthfully indicate the geometry in the region under consideration. Accordingly, we have to make sure prior to performing our measurements that the rods used are of the right kind, i.e., that they are *rigid*. But how can we verify that our rods are actually rigid; put the other way around, how can we exclude that they are subject to position or time dependent deformations?

Two sorts of relevant deformations are conceivable. The first one is produced by forces that act differently on chemically distinct substances. Such forces are called *differential* by Reichenbach. Temperature is a case in point. The length of rods is altered by temperature changes, and the magnitude of this alteration varies with the rods' material. The functioning of gas and liquid thermometers is based on the fact that a rise in temperature expands the respective indicator substance (i.e., mercury, alcohol and the like) to a larger degree than it expands the surrounding glass tube. Because of their substance-specific nature, differential forces can easily be detected and evaluated; their influence can accordingly be corrected.[20]

Things are not that simple, however, with respect to the second kind of deformation. Imagine a force which acts alike on all materials. Forces of this kind are called *universal* by Reichenbach. It is not possible to empirically evaluate the intensity (or the mere presence) of a universal force directly. The reason is that lengths can only be compared locally; so if two locally congruent rods are transported along different paths and are again found to be locally congruent, it cannot be ruled out empirically that *both* rods have changed their length during transport in a universal fashion. It is impossible to directly compare lengths at distinct locations.

Recourse to optical methods of length comparison is of no help. For such methods evidently rely on assumptions about the propagation of light, and it is an easy matter to introduce an appropriate deviation of light rays from the straight path under the action of a universal force (cf. IV.2.§2–3). Accordingly, one is free to explain away the results of optical measurements by supposing that these procedures are subject to the same influences as transport experiments. Without prior commitment to a theory about the behavior of rods under transportation and about light propagation, it is impossible to obtain data suited to establish uniquely a relation of distant congruence. Reichenbach's conclusion is that distant congruence has to be *defined* instead (cf. Reichenbach 1928, 10–14).

This definition is given by stipulating what is to qualify as rigid rods. Such rods are identified by means of a *coordinative definition.* This amounts to specifying an operationally applicable criterion for their identification. Reichenbach's proposal for this coordinative definition is as follows. Rigid rods are solid rods which meet two conditions: First, the influence of all possibly present differential forces is corrected; second, all universal forces are set equal to zero (cf. Reichenbach 1928, 22). Whereas differential effects are construed as distortions and consequently rectified, universal deformations are excluded by definitional *fiat*.

Only after a coordinative definition of this or similar kind has been given, it is sensible to ask for the structure of physical geometry. Every empirical investigation into the prevailing geometric relations is dependent upon and relative to the standard of rigidity chosen. The adoption of such a standard, and the corresponding decision about the presence or absence of universal forces, is not only necessary for fixing physical geometry, it is sufficient as well. Thus, there are two claims involved here. First, after a stipulation about universal forces has been accepted, the metric (together with all geometric relations derivable on its basis) is determined unambiguously. Second, it is recommended that we make this stipulation such that universal forces vanish.

One now gets a hint as to where Reichenbach's approach will end up eventually. It ends up with gravitation; more specifically, with Einstein's theory of gravitation. For it is gravitation that fits Reichenbach's description of a universal force; this description is tailored so that gravitation comes out as a universal force. The weak principle of equivalence expresses precisely the independence of gravitational forces from the internal constitution of bodies; and this boils down to the fact that gravitation acts on all materials alike – as is constitutive of a universal force. What is it supposed to mean, then, that gravitation be set equal to zero? It means that it is not to be construed as a force that demands correction. The decision to abjure universal forces amounts to interpreting the effects of gravitation not as a deformation, but as veridical indication of the underlying space-time structure. This decision thus amounts to leaving gravitational influences uncorrected. This is how Reichenbach's rule leads to geometrized gravity (cf. Reichenbach 1928, 256). So, the whole point of the argument is to provide a justification for General Relativity Theory in its familiar shape. The conventionality thesis entails that a large number of conceptually different, but empirically equivalent accounts of physical geometry are conceivable. For instance, there are no facts militating against accounting for gravitation by a Newton-type universal force operating in flat space-time.[21] Consequently, Einstein's particular variant cannot be distinguished on empirical grounds. Singling out this variant as the appropriate one thus necessitates recourse to non-empirical criteria. Reichenbach's rule furnishes such a criterion; in its light Einstein's approach qualifies as superior.[22]

Accordingly, standard General Relativity is not thought to be empirically privileged as compared to universal-force approaches. General Relativity only scores better on criteria such as simplicity and convenience; it is merely distinguished by its "descriptive simplicity." But this advantage has nothing to do with truth. It constitutes a virtue analogous to that of metric units. Metric units are easier to handle than alternative ones (such as inches and ounces); but obviously a meter does not fit reality better than an inch does. Adopting one geometric account at the expense of an alternative one is ultimately a matter of convention.[23]

This is not intended to mean, however, that as regards the facts geometry is totally arbitrary. For instance, it is a *fact* that local congruence is preserved under transportation. If two rods whose lengths have come out equal in direct comparison are transported along different paths to the same location, they are again found to be of equal length (after the correction of possibly present differential distortions). It is by virtue of this fact that the above-given definition of a rigid rod is unambiguous in the first place. Moreover, the relation between the adoption of a standard of rigidity and the ensuing geometry is factual as well. It is a *fact* of the physical world – undreamed-of before Einstein – that setting universal forces equal to zero makes physical geometry non-Euclidean in character (cf. Reichenbach 1928, 16–17, 23–24, 37).

§ 2. It will be helpful to become a bit more familiar with the epistemic status and the physical nature of universal forces. To begin with, it should be emphasized that Reichenbach's conventionality argument in no way boils down to the claim of traditional epistemology, that an overnight expansion of all lengths by the same factor would go unnoticed. This claim is simply mistaken; the expansion would certainly have observable consequences. First, a doubling of all lengths would entail that all measured velocities, including natural constants as the velocity of light, would assume half their familiar magnitudes. Only by the additional adjustment of other, related quantities can these effects possibly be explained away (cf. Putnam 1963, 110; Hempel 1966, 97). Second, doubling all lengths leads to a corresponding doubling of the ratio between volume and surface of all bodies, and this effect would become manifest in alterations of their thermal behavior. For instance, after the expansion, all bodies would cool off more slowly than they did before. Third, such a general expansion would induce discrepancies among methods for measuring durations that were formerly in agreement. Namely, whereas doubling the dimensions of a watch operating with a balance spring doubles its period of oscillation (cf. Ohanian 1977, 905), it increases a simple pendulum's period of oscillation only by a factor $\sqrt{2}$. Consequently, a balance-spring clock and a pendulum clock running at equal rates would run at different rates after a general length expansion. The conclusion is that the traditional argument in favor of the relativity of lengths is unsound.[24]

But Reichenbach's conventionality thesis is by no means based on a simple undetectability claim for deformations induced by universal forces. Quite the contrary. Such deformations are admittedly detectable. First, given a particular geometry the presence or absence of universal forces clearly makes an empirical difference. If a flat background metric is assumed, the presence of a gravitational field certainly has observable consequences. The point is rather that the origin of these effects is not directly amenable to experience. They can be attributed either to a changed geometry or to a force field.

Second, independent of any presupposed geometry, a gradient in a universal force-field may manifest itself in the occurrence of relative deformations. If the field intensity is different at different locations, the forces it exerts on the different parts of a body – provided it is large enough to be affected by the non-uniformity of the field – will be different as well. Since a stronger force induces a greater length alteration, a non-uniform field tends to distort the shape of an extended body, and this effect is, of course, easily accessible empirically (cf. Reichenbach 1928, 24–26). Clearly, what Reichenbach has in mind here are tidal forces (cf. IV.3.§7).

Third, the material-independence of a universal force does not imply the material-independence of the induced deformation. Rather, the deformation is the joint effect of that force and the intensity of cohesive forces that tend to maintain the body's shape. Since these cohesive forces vary with the material under consideration, the resulting deformation does so as well. After all, gravity obviously deforms different substances to a different degree. Suspending a given weight from an elastic spring induces a larger deformation in that spring than suspending it from a concrete ceiling does in the concrete. Due to different internal forces the response of different media to the same external force is different.

Reichenbach clearly recognizes, and in fact emphasizes, that universal forces do not preserve length relations under all circumstances. Setting universal forces equal to zero does not amount to the recommendation to disregard theoretically what can never become manifest empirically. As he puts it:

> Forces *destroying coincidences* must also be set equal to zero, if they satisfy the properties of universal forces [...] Our concept of universal force is thus more general and contains the concept of the coincidence-preserving force as a special case (Reichenbach 1928, 27).

Setting such empirically accessible forces equal to zero means, then, to attribute their effects to space-time structure. It means, not to interpret them as perturbations and not to rectify their influence.

§ 3. Now the time has come to connect the conventionalist scenario to the problems of testing Einstein–Feigl complete theories or hypotheses. The conventionalist point is that there exist various alternatives for coping theoretically with observed length relations. These alternatives are to be sorted into two different types of approaches. On the one hand, one may interpret possibly occurring substance-unspecific effects as real alterations of the rods' lengths as produced by a universal force. In that case such rods are thought not to indicate correctly the prevailing geometric relations. On the other hand, one may geometrize away the universal force field by construing the deformations as arising from the underlying space-time structure. In that case these rods are regarded as still truthfully indicating the geometric relations.

The two approaches differ in what they consider as the pure state and what as a distortion. The universal force account interprets gravitational influences as a distortion of the lengths of the rods and insists in correcting for the effects of these influences. The geometric approach, by contrast, regards (infinitesimal) rods in a gravitational field as undistorted, i.e., as rigid, and consequently denies the need for any rectification of results obtained that way.

The crucial point can be brought out more concisely using an equation given by Reichenbach. Let the true spatial geometry be $g_{ab}(a, b) = 1, 2, 3)$ and assume that the measuring results suggest a spatial geometry $^*g_{ab}$ different from g_{ab}. Harmony can now easily be regained by introducing a spatial universal force U_{ab}. It is claimed, then, that the geometric relations are described by the equation:

$$(5) \qquad {}^*g_{ab} + U_{ab} = g_{ab}$$

That is, the measured raw-data have to be corrected by appealing to the universal force U_{ab} (cf. Reichenbach 1928, 33).

Now the connection with completeness leaps to the eye. For both the undistorted metric and the distorting force are exclusively specified by Equation (5). We are presented with two quantities and only one relevant equation, and this situation is constitutive of a *Reichenbach loop* as introduced in II.3.§1. In fact, it is this example which originally led me to coin the term. In this sense we are now dealing with *the* paradigmatic Reichenbach loop.

Reichenbach loops, it will be recalled, are a possible complication of potential completeness on the hypothesis-level. The latter is characterized by the fact that all empirical indicators of a quantity are specified by one single hypothesis (cf. II.1.§4). In Reichenbach loops, in particular, it is two quantities that can only be evaluated by resorting to one hypothesis, with the result that a reciprocal dependence between these two quantities arises. Either one can only be ascertained if the complementary one is already given. We are thus caught in a circularity that endangers the unambiguous determination of the two quantities involved. In the case under consideration this circularity concerns the true metric and the universal force field. Even by presupposing the truth of Equation (5), we are at a loss to infer definite values for either of these two quantities from the geometric data. It is thus always possible to satisfy Equation (5) whatever the evidence may be. Because of this guaranteed fit, Equation (5) is not bootstrap-testable, i.e., not testable by definite instantiation. Bootstrap testability, it will be recalled, requires that instantiations could possibly turn up which would contravene the hypothesis at issue (cf. II.4.§1).

Accordingly, the argument advanced in favor of the conventionality physical geometry can be reconstructed to the effect that the empirical determination of geometry is thwarted by a Reichenbach loop which renders a test by instantiation impossible. The conventionality thesis thus bears witness to the fact that in establishing physical geometry we are faced with a non-Duhemian test problem. The occurrence of such a test problem prevents us from actually reaching completeness on the hypothesis level.

The problem can be restated in different terms. As described in II.5.§3, some Reichenbach loops can be construed as corrective loops and then issue in self-referential distortions. Performing the correction of a self-referential distortion demands reference to an evaluation of the same quantity whose measurement is to be corrected. It is easily recognized that all features of that scenario are realized in the present case. Universal forces are self-referential distortions of metric measurements. The inference from the distorted $^*g_{ab}$ to the undistorted g_{ab} requires recourse to the corrective force. This force is accessible in principle through a comparison between the pure state and the distorted evidence. But such a comparison necessarily has to appeal to Equation (5): $U_{ab} = g_{ab} -^* g_{ab}$. That is, in order to arrive at g_{ab} we are enforced to apply a correction procedure to $^*g_{ab}$ which essentially makes use of U_{ab}. But the evaluation of U_{ab} in turn demands reference to g_{ab}. The influence of a universal force on the true metric can only be corrected by resorting in turn to the true metric. Again, the threat of a vicious circle becomes obvious.

§ 4. The natural move now is trying to counter the circularity threat by bringing to bear the two chief strategies especially designed for handling Reichenbach loops. Application of non-instantiative testing or external fixing should be suitable for restoring testability. But in fact, it is not. Non-instantiative testing demands that an empirically testable consequence of the hypothesis in question be instantiated, and that this instantiated consequence be such that it allows for a discrimination among alternative variants of the hypothesis (selective entailment) (cf. II.3.§5). We can see at once that this latter condition is not met in the case at hand. By virtue of the empirical equivalence of the various geometry-force conjunctions, their instantiated consequences are in perfect agreement with one another and thus obviously fail to discriminate among these conjunctions.

It is not difficult to identify the origin of this failure. In all successful applications of non-instantiative testing we could rely on constraints effectively restricting the magnitudes of the quantities which are underdetermined by the corresponding hypothesis alone. Typically, such helpful constraints have been constancy constraints or additivity constraints (cf. II.3.§2–4). But nothing of that sort is available in the present problem-situation. There are no separate constraints for geometry or force shared by alternative approaches. Consider, for example, a gravitational field of spatiotemporally changing intensity. In that case we look in vain for a constancy constraint that is generally agreed upon. The geometric approach considers the force to be constant (namely, zero) and the geometry to be variable, whereas the paradigmatic representative of the force approach regards geometry as being unchanged (namely, as being described by the flat Minkowski metric) and the force as variable. Since appropriate constraints are missing, non-instantiative testing breaks down (as in the analogous case discussed in II.3.§4). Equation (5) thus cannot be tested non-instantiatively.

The second test option is external fixing. On that approach we must ask whether there are any additional hypotheses suited to establish definite values for one of the quantities involved. Well, there are indeed, but only in a Pickwickian sense. A hypothesis in point is Reichenbach's rule, for instance. Application of this rule fixes universal forces to a definite value: zero (see § 1); it thus makes the loop equation (5) collapse into an unproblematic identity between measured and true geometry. Hence, testability is restored, but it is restored merely relative to the adoption of Reichenbach's rule.[25] Passing from testability and confirmation relative to an auxiliary to credibility (i.e., to confirmation *simpliciter*) demands that this auxiliary possess some empirical credentials on its own.[26] And Reichenbach's rule obviously fares rather poorly in this respect.

Couched in terms of distortions, the resolution of a self-referential distortion circularity requires that the undistorted case be identifiable. In the geometric scenario this is perfectly possible – provided that a prior decision about what is to count as a distortion has been made. But it is precisely this decision that the argument among the factions is all about. It is the nature of the pure case that is under debate (see § 3). In sum, there are no hypotheses that could possibly fix one of the quantities at hand *and* are generally agreed upon on empirical grounds. This is why external fixing is no avail here either.

The conclusion is that the circularity generated by the universal force loop cannot be remedied by applying our standard antidotes. Regarding physical geometry, the non-Duhemian test problem is here to stay. The reason is that neither the universal force nor the metric resulting from it, as distinct from the measured metric, are sufficiently integrated into the body of physical laws. The geometric Reichenbach loop is isolated theoretically. It is this isolation that prevents us from deploying successfully the familiar techniques. This time the circle is real and vicious.

But if the Reichenbach loop inherent in the determination of physical geometry is really intractable theoretically, why not do away with it? Why not apply the therapy sketched in II.4.§7 and simply amputate the loop altogether. In a sense, this is what Reichenbach proposes: Let's forget about the correction term U_{ab} and proceed on the basis of the measured $^*g_{ab}$ alone. But the drawback of that simple solution is that, intuitively, it certainly makes a difference whether gravitation is a physical force or incorporated into physical geometry. It certainly makes a difference whether the force account or the geometric account of the falling apple (cf. IV.3.§5) is more accurate. Other than the examples of rightly amputated loops in II.4.§7, and contrary to the conventionalist contention, this kind of difference is not adequately captured by comparing it to choices of units. Decisions about the machinery underlying gravitational effects should not be treated on a par with decisions about the most convenient system of units. These considerations suggest that the Reichenbach-loop inherent in the empirical determination of physical geometry should not be rejected as spurious.[27]

§ 5. Reichenbach's treatment suffers from the fact that he considers spatial and temporal geometry separately (for the latter see § 6). Universal forces are primarily conceived as spatial or three-forces (as is expressed in Equation (5) in § 3). This separate treatment does violence to the essentially four-dimensional spirit of General Relativity. In the framework of that theory, spatiotemporal geometry does not always split up into a spatial and a temporal geometry, and when it does, the split is frame-dependent. There are situations, to be sure, in which a spatial geometry and, correspondingly, three-forces can be introduced in a natural fashion. Among them is the situation of a rod described in its rest frame (cf. Earman/Friedman 1973, 355). But it is awkward to focus the argument on distinguished cases of this kind.

Actually, Reichenbach is fully aware that it is four-forces and not three-forces that matter. It is the former, and not the latter, that are geometrized in the first place.[28] But Reichenbach's actual line of argumentation does not take this insight into due account, and precisely for this reason his discussion starts off from the wrong foot. His point can be brought more into keeping with the four-dimensional approach of General Relativity by translating it into trajectory language. This has been done by Glymour. So, let's see what form Reichenbach's argument takes if we consider the motions of particles rather than the lengths of rods. As a result of the equality of inertial and gravitational mass, a gravitational field accelerates all bodies in an equal measure irrespective of their internal constitution (cf. IV.3.§1). Consequently, gravitation acts on the paths of all particles alike; it deforms the particle trajectories in a universal fashion. It follows that the trajectory of a particle in a gravitational field is independent of that particle's mass, chemical nature and so forth.

This consideration can be condensed into the following equation of motion for a particle moving in a gravitational field. The gravitational field strength (i.e., the force per unit mass) is denoted by U^i, and the gist of the foregoing discussion is captured by equating this field quantity with the particle acceleration. Accelerations are to be expressed in the generally covariant manner as indicated on the left-hand side of Equation (4) (cf. IV.3.§4). The equation of motion is then:

$$\text{(6)} \qquad \frac{\mathrm{d}^2 x^i}{\mathrm{d}\tau^2} + \Gamma^i_{jk} \frac{\mathrm{d}x^j}{\mathrm{d}\tau} \frac{\mathrm{d}x^k}{\mathrm{d}\tau} = U^i$$

The universal nature of gravitation is expressed by the fact that no particle parameters appear on the left-hand side of Equation (6). The acceleration conferred by a gravitational field is independent of the respective particle's constitution.

On this approach, the motion of a particle under the exclusive action of a gravitational field displays a non-vanishing acceleration. The particle trajectory is diverted from the corresponding geodesic through the action of the field. The particle moves non-inertially. Now the trick is to tamper with the law of inertia. We introduce a changed geodesic structure to the effect that the particle now moves along a geodesic. That is, with respect to the altered geodesic structure the particle moves unaccelerated. The geodesic structure is characterized by the affine connection (cf. IV.3.§4). So, what we do is change the former connection Γ^i_{jk} into the new connection ${}^*\Gamma^i_{jk}$ with the result that the previously accelerated motion becomes inertial. Equation (6) accordingly goes over into:

$$(7) \qquad \frac{\mathrm{d}^2 x^i}{\mathrm{d}\tau^2} +^* \Gamma^i_{jk} \frac{\mathrm{d}x^j}{\mathrm{d}\tau}\frac{\mathrm{d}x^k}{\mathrm{d}\tau} = 0$$

If expressed using of the abbreviations $t_j = (\mathrm{d}\tau)/(\mathrm{d}x^j)$ and $t_k = (\mathrm{d}\tau)/(\mathrm{d}x^k)$, the new connection, obtained from combining Equation (6) and Equation (7), has the form:

$$(8) \qquad {}^*\Gamma^i_{jk} = U^i\, t_j t_k - \Gamma^i_{jk}$$

By means of this maneuver the universal field is incorporated into the geodesic structure of space-time (cf. Glymour 1977, 245, Glymour 1980, 366).

Consequently, what previously appeared as a curved trajectory is now construed as motion along the straightest line possible. And this is accomplished by a change of mind about which lines are to be regarded as straightest and a corresponding reassessment of the underlying space-time structure. It is clear that this constitutes a workable option only if the action of the field is indeed independent of the nature of the particle. Otherwise we would end up with a geometry different for different particles which would appear somewhat artificial.

So we are indeed presented here with another implementation of the strategy to geometrize away a universal force. The universal force account and its geometric analog are empirically equivalent; Equation (6) and Equation (7) entail the same observable consequences. And again, it is the presence of a Reichenbach-loop that is responsible for the inability to distinguish empirically between the two approaches. The logical structure of Equation (8) agrees with that of Equation (5) (see § 3) which describes the influence of spatial universal forces on spatial length relations. There is a reciprocal dependence between the geodesic structure one assumes and the universal effect one appeals to, with the result that both these quantities are underdetermined by experience. Accordingly, the concept of force-free motion is no more uniquely specified empirically than the concept of rigidity. In both cases the facts leave room for alternative approaches.

Again, the conclusion is that completeness on the hypothesis level may entail a breakdown of testability. Equation (8) is as untestable as Equation (5). The Reichenbach loop is left undissolved by switching from a spatial to a spatiotemporal perspective. And since there are no other, independent options for an empirical access to either the true geometry or the distorting universal effect, we are faced with a potential case of strong Einstein–Feigl completeness on the hypothesis level. It is this feature that lies at the bottom of the conventionality claim regarding physical geometry.

§ 6. From space through space-time to time. In the empiricist and conventionalist tradition the problem of determining durations was conceived in close analogy to the problem of singling out empirically lengths of equal magnitudes. As Reichenbach argues, equal time intervals can only be established by resorting to periodic processes. But not all such processes are equally suited for that purpose. We need processes that go on uniformly. But the uniformity of a given process can only be ascertained by comparing its course to some other process known in advance to proceed uniformly. The infinite regress lurking here is quite analogous to the one that emerges with respect to spatial geometry (see § 1). In particular, a temporal universal force may increase or decrease the duration of all periodic processes in a like manner. Reichenbach's conclusion is that we need a coordinative definition for temporal congruence as well, and that this definition should conveniently be made so that the laws of physics assume the simplest shape possible.[29]

In fact, Reichenbach's conclusion echoes an earlier contention of Poincaré's:

> Time is to be defined in such a way that the mechanical equations are as simple as possible. In other words, there is no method for measuring time that is more *true* than another one; the one that is generally adopted is only more convenient (Poincaré 1898, 6, translation mine).

The same view is expressed some decades later by Misner–Thorne–Wheeler as follows: "Time is defined so that motion looks simple!" (Misner/Thorne/Wheeler 1973, 23).

There is, however, one more problem involved in the empirical determination of temporal relations; namely, the problem of specifying a relation of *distant simultaneity* that can be justified on empirical grounds. This problem is more properly to be attributed to Special rather than General Relativity. Its importance derives from the fact that it gives rise to an additional procedure for coping with the test problems inherent in Reichenbach loops. The present subsection is devoted to the presentation of the problem while its partial solution is discussed in IV.5.§2–5.

In order to establish a relation of distant simultaneity, we have to synchronize distant clocks. The most convenient way to do this seems to be to use signals, that is, light rays. If a light signal arrives at a clock at time t_2, we can calculate the time t_1 at which it was emitted at a distant source. These two values allow us to infer a synchrony relation between distant clocks. But the procedure demands that the velocity of light is known, and it is precisely this proviso which we run into trouble with. For the measurement of the velocity of light requires that we are able to determine the amount of time a light signal takes to traverse a known distance. And the latter determination obviously presupposes that distant synchronized clocks are already available. We are thus trapped in a circle.

But one may resort to a different experimental set up. Consider a mirror that is situated at known distance from a light source. At time t_1, a light pulse is emitted in the direction of the mirror. It is reflected there and sent back to its origin, where its return time t_3 is recorded. Now we may plausibly assume that the time t_2 of distant reflection is just half the overall travel time. That is: $t_2 = t_1 + \frac{1}{2}(t_3 - t_1)$. Note that we have employed only one clock located at the origin so that the problem of using synchronized clocks does not show up here.

But this easy solution suffers from a serious drawback. It is presupposed that the forward and backward velocity of light are equal. What we actually measure is merely the *average* velocity or *two-way* velocity, and what we need in order to arrive at the above given value for t_2 is the *one-way* velocity. The method employed proceeds on the assumption that the light propagation is *isotropic*; this assumption is consequently not tested by the experiment. On the basis of the data we are only allowed to conclude that reflection occurs *at some time* between emission and return. The experiment thus only licenses the following equation for t_2:

(9) $$t_2 = t_1 + \varepsilon(t_3 - t_1) \quad 0 < \varepsilon < 1$$

That is, we do not run into any contradictions with the empirical evidence if we claim that light travels faster in the forward direction (or the backward direction, as we prefer) – provided that the two-way velocity is kept unchanged. Accordingly, we are free to assert that the event of reflection does not cut the round trip time into two equal halves. Within the just-stated interval, each ε-value has equal status; it is up to us to pick one. Accordingly, distant simultaneity, as characterized by a certain choice of ε, cannot be established empirically but has to be stipulated by a *coordinative definition*. Distant simultaneity is a matter of *convention*. The usual choice of $\varepsilon = \frac{1}{2}$ (i.e., isotropy of light propagation) is more convenient than alternative values but none of them is privileged empirically and thus "more true" than any other.[30]

§ 7 The conventionality claim regarding distant simultaneity grows out of the observation that distant simultaneity is underdetermined by the data under consideration. And again, the untestability can be reconstructed as being the result of strong Einstein–Feigl completeness on the hypothesis level. A Reichenbach loop is inherent in the empirical determination of distant simultaneity. The relevant hypothesis is Equation (9). Two quantities figure in that equation which are not immediately amenable to experience, namely, ε, representing a relation of distant simultaneity, and t_2, expressing the one-way velocity of light. In all non-standard synchronizations ($\varepsilon \neq \frac{1}{2}$), the one-way velocity (c_{ow}) is direction-dependent[31] since the back and fro velocities are supposed to differ.

Since ε and c_{ow} are determined by only one equation, there arises a reciprocal dependence between their respective evaluations. Assigning a value to ε, i.e., distinguishing a relation of distant simultaneity, requires that c_{ow}, i.e., the one-way velocity of light, has been evaluated. But in order to justify the latter evaluation one has to rely on a specific ε-value. The ensuing circularity is expressed by Reichenbach as follows.

> Thus we are faced with a circular argument. To determine the simultaneity of distant events we need to know a velocity, and to measure a velocity we require knowledge of the simultaneity of distant events. The occurrence of this circularity proves that simultaneity is not a matter of knowledge, but of a coordinative definition, since the logical circle shows that a knowledge of simultaneity is impossible in principle (Reichenbach 1928, 126–127).

So, the only way out appears to be to arbitrarily pick a value for one of these quantities and to correspondingly adjust the other one. Considering exclusively the empirical consequences each choice is equally legitimate.

The conventionality of physical geometry, as conceived by Reichenbach, is conditional on general epistemic limitations. The conventionality of distant simultaneity, by contrast, is construed as being based on more specific physical restrictions imposed by Special Relativity Theory (cf. Grünbaum 1973, 346). In particular, it is based on the fact that the maximum velocity of signal transmission is finite. In a Newtonian world, in which the influence of gravitation spreads infinitely fast, no conventionality of distant simultaneity would arise. This example makes it clear, in addition, that in order for the conventionality argument to go through, a second condition has to be satisfied. There must not exist relevant procedures that are not afflicted with the circularity described and thus fit to dissolve the underdetermination of distant simultaneity by experience. Reichenbach argues that there aren't indeed any such procedures and that, in particular, even our background knowledge is insufficient to give preference to any specific ε-value.[32] Because of the alleged non-availability of alternatives, the determination of either one of the quantities involved *necessarily* has to resort to Equation (9). And this is why we are presented with a potential case of *strong* completeness on the hypothesis level.

The Reichenbach loop inherent in the empirical determination of distant simultaneity may as well be described as self-referential distortion (just as the analogous problem of measuring the true metric (see § 3)). On that construal, the anisotropy of light propagation (i.e., the direction-dependence of c_{ow}) is viewed as a perturbation intruding into every measurement of distant simultaneity. Since the presence of this perturbation has to be reckoned with, we are prevented from simply basing our considerations on the undistorted case $\varepsilon = \frac{1}{2}$. But the magnitude of the distorting effect can only be calculated by measuring one-way velocities in different directions which in turn presupposes that a relation of distant simultaneity has been established beforehand. That is, rectifying the influence of a possible anisotropy on distant simultaneity necessitates recourse to a relation of distant simultaneity. And this feature instantiates the self-referential-distortion pattern.

Is it possible to repair this flaw? First, it is of no avail to invoke non-instantiative testing. The reason is the same as in the analogous case of physical geometry (see § 4). There exist no conditions effectively constraining admissible values of the relevant quantities. For instance, c_{ow} is thought to be variable in non-standard synchronizations but constant in the standard approach. Put more generally, there are no constraints for c_{ow} or ε that are strict enough to allow for the derivation of testable relations sensitive to different concrete versions of Equation (9). Since all these versions are empirically equivalent, their instantiated consequences coincide. And provided that there are really no further procedures, not relying on Equation (9) and suitable for establishing a definite relation of distant simultaneity, external fixing fails as well. It is this purported lack of alternatives which constitutes the subject of the next section.

5. THE COMPLETENESS OF SPECIAL RELATIVITY REGARDING DISTANT SIMULTANEITY

§ 1. In the present section I explore the options for overcoming the circularity inherent in the empirical determination of distant simultaneity. Since the circle actually has the structure of a Reichenbach loop, this task amounts to exploring the options for coping with problems which potential completeness on the hypothesis level may entail. True, specifying a simultaneity relation is primarily a problem of definite instantiation rather than testing. Even so, there is a large common ground between both problems, and it will be seen that the two chief approaches to tackling Reichenbach loops prove helpful with respect to the instantiation problem as well.

It deserves emphasis, to begin with, that there is indeed no direct empirical access to the one-way velocity of light. This has been confirmed by Wesley Salmon, for instance, through an extensive survey of purported measuring procedures for that quantity. It turns out that every such method presupposes and tacitly employs a convention about distant simultaneity.[33] Correspondingly, the reciprocal dependence between these two quantities cannot be avoided or removed by applying sophisticated technical means. And as I argued in IV.4.§7, the vicious circularity suggested by this reciprocal dependence cannot be circumvented or defused by having recourse to alternative conceptual strategies such as non-instantiative testing. So we are left with the conclusion that the circularity involved in the use of light signals for synchronizing distant clocks is irremediable. Light signals provide no means for distinguishing empirically a unique relation of distant simultaneity.

This result does not rule out, however, that there exist other methods that prove successful where signal synchronization fails. Of particular interest in this respect is the *transport of clocks*. After all, the traditional method for sending a message is relying on a courier service. The electric telegraph was preceded by the stage coach. In a sense, clock transport is of course just another way of sending a signal. Still, employing a less advanced and more traditional technology pays off on occasion.

Accordingly, a promising procedure for synchronizing distant clocks via clock transport seems to be the following. We have two distant clocks C_A and C_B, and we locally synchronize a third clock C_C with C_A. That is, these two clocks have been adjusted by direct comparison so as to yield identical time readings. Subsequently, we transport C_C to the distant clock C_B, and we set the latter according to the reading of the former. As a result of these two local synchronizations, C_A and C_B are in distant synchrony to one another – or so it seems.

But unfortunately, this simple solution does not work. For there is a problem with relativistic time dilation. Special Relativity entails that a moved clock lags behind its counterpart at rest. The time lag Δt depends on the transport velocity v and the length of the path s along which it is transported. If standard synchrony ($\varepsilon = \frac{1}{2}$) is assumed, the moved clock is slowed relative to the one at rest by the amount:

$$(10) \qquad \Delta t = \frac{1}{2}\frac{vs}{c^2}$$

In fact, this expression is only a close approximation to the exact formula, but it suffices for our purpose. Equation (10), along with its exact counterpart, was derived by Einstein already in his celebrated 1905 paper which laid the foundation for Special Relativity Theory (cf. Einstein 1905, 48–49).

Relativistic time dilation brings the following difficulty in its train. Given a fixed distance s between the two clocks C_A and C_B, the time reading of C_C, initially in local synchrony with C_A, varies upon its arrival at C_B with the transport velocity of C_C along its path. The results of transport synchrony depend, in general, on the particulars of clock transportation and thus lead to inconsistent values. A transported clock ceases to be self-synchronous. It was this unhappy peculiarity that led Reichenbach to abandon transport synchrony and to focus on signal synchrony with all its (meanwhile notorious) difficulties (cf. Reichenbach 1928, 133).

Let's look at this problem more closely. In the framework of the approach pursued here, time dilation can be interpreted as a distorting influence on the clock's running.[34] As a result of this distortion, the raw-readings of differently transported clocks are in disagreement and thus unsuitable as an empirical basis for a unique simultaneity relation. But a distortion of this kind should be correctable. After all, Equation (10) provides a quantitative evaluation of the distorting effect. And as a glance at this equation reveals, it is the transport velocity that constitutes the crucial factor in this evaluation. So we are led to the following proposal for a correction procedure. We measure the transport velocity, calculate from it the induced time lag via Equation (10), and use the value obtained for the correction of the reading of C_C upon its arrival at C_B. It is this rectified C_C reading that is supposed to furnish the basis of transport synchronization.

But this smooth plot glosses over a lacuna which seriously threatens its practical implementation. For what we need is the *one-way transport velocity*. After all, C_C is transported one way from C_A to C_B. But how do we measure this one-way transport velocity? Such a measurement obviously presupposes that we have already distant, synchronized clocks at hand. It is clear, moreover, that the one-way transport velocity cannot be inferred from the two-way or round trip velocity since in the transport scenario the possibility of different back and forth velocities is even more conspicuous than in the light signal scenario. So, the one-way transport velocity is no more amenable to empirical determination prior to a simultaneity convention than is the one-way velocity of light.

In fact, the unmeasurability of the one-way velocity of light entails that *no* one-way velocity whatever is empirically accessible without appeal to a relation of distant simultaneity. For suppose we know the one-way velocity of some object or causal process. In that case we may send off simultaneously this entity along with a beam of light from a common origin in a common direction. At a known distance from the origin we record the arrival times of the light beam and the reference entity. Then we can easily calculate the one-way speed of the former from the one-way speed of the latter, the difference in their arrival times, and the distance covered. Only local synchronization is involved in this procedure. The viability of the conventionality claim as to the one-way velocity of light requires, therefore, that no other one-way velocity can be determined without presupposing a relation of distant simultaneity (cf. Salmon 1977, 270–271; see also Winnie 1970, 84–86). The just-sketched problem inherent in transport synchronization thus constitutes a particular instance of this comprehensive unmeasurability claim.

This consideration makes it clear that we again get stuck in a Reichenbach loop. There is a reciprocal dependence between measuring the one-way transport velocity and determining distant simultaneity. This reciprocal dependence appears to vitiate the unambiguous evaluation of either quantity. The circle involved here thus exhibits exactly the same logical pattern as the one present in the light signal scenario. Accordingly, it can again be couched in terms of the concept of self-referential distortions (cf. IV.4.§7). The determination of distant simultaneity by means of a transported clock is perturbed by the influence of the transport velocity on the clock's running. The rectification of this perturbation demands measurement of the one-way transport velocity which in turn requires reference to a relation of distant simultaneity. Relativistic time dilation is thus a self-referential distortion of distant-simultaneity measurements. And it seems that this self-referential distortion gives rise to serious trouble regarding the unambiguous instantiation of either quantity involved.

§ 2. However, it is possible to refine the transport method in such a way that it indeed provides a basis for singling out empirically a unique relation of distant simultaneity. In other words, the Reichenbach loop inherent in this determination can actually be overcome. Let's now see, first, how this feat can be performed and, second, how this procedure is to be understood in terms of completeness.

As Brian Ellis and Peter Bowman observed, following an earlier suggestion of Percy Bridgman, the influence of relativistic time dilation decreases as the velocity of the transported clocks is diminished. In the limiting case of infinitely slow transport, time dilation vanishes. Judging from Special Relativity, therefore, an infinitely slowly transported clock establishes a unique relation of distant simultaneity that coincides with standard synchrony ($\varepsilon = \frac{1}{2}$). But the problem is, it will be recalled, that we are prevented from evaluating *any* transport velocity, and this unfortunate peculiarity seems to preclude any inference to infinitely slow transport.

But in fact this problem can be solved by invoking "self-measured velocities", as Bridgman called them (cf. Bridgman 1962, 65). Referring back to the transport scenario with two stationary clocks C_A and C_B and a transported clock C_C, we define the self-measured velocity of C_C

as follows. This quantity is given by the ratio of the distance between the two stationary clocks, as measured in their rest frame, and the transit time, as measured by the moved clock. Hence, the self-measured velocity combines values taken from two different inertial frames and thus constitutes "truly a bastard concept" (Salmon 1977, 265). But as is the case for many bastard breeds, this particular one as well will prove robust.

It is clear, first, that the self-measured velocity of C_C does *not*, in general, agree with the velocity one would have obtained for C_C by using two stationary clocks in standard synchrony. This disagreement signals relativistic time dilation. The crucial point is, however, that it can be shown using Special Relativity that if the self-measured velocity v_s converges to zero, the standard velocity v does so as well. Special Relativity implies, namely, that the following relation between these two quantities holds:

$$(11) \qquad v = v_s \left(1 + \frac{v_s^2}{c^2}\right)^{-1/2}$$

Judging from this equation, $v_s \rightarrow 0$ implies $v \rightarrow 0$. And this means that transport with infinitely small self-measured velocity leads to the same result as transport with infinitely small standard velocity. It follows from Equation (10), however, that in the latter case the dilational influence is negligible, and this entails that it is likewise negligible in the former case. This is tantamount to saying that synchronization by means of infinitely slow clock transport establishes standard synchrony, i.e., $\varepsilon = \frac{1}{2}$, between the distant stationary clocks C_A and C_B.[35]

The last point to be addressed is then: How is infinitely slow transport to be realized? After all, transport with vanishing transport velocity doesn't strike us as an overly clear notion. While it is true that this type of transport cannot be realized directly, it can be approached by employing a procedure of *successive approximation*. That is, we transport several clocks (or the same clock several times) with decreasing self-measured velocity from C_A to C_B and extrapolate the results obtained to the limiting case of zero velocity. Although this latter case is thus not amenable to direct scrutiny, its characteristics can be inferred from a series of repeated transport experiments. The conclusion is, first, that Special Relativity allows for a non-circular determination of distant simultaneity and that, second, the simultaneity relation ensuing from this determination is precisely standard synchrony.

§ 3. This measurement procedure is indeed suitable for providing a distinguished relation of distant simultaneity. So let's recapitulate its essential steps in order to understand what makes the relevant Reichenbach loop disappear. This loop was generated by the reciprocal dependence between distant simultaneity and one-way transport velocity. Hence it seemed that the reliable evaluation of either quantity is conditional on the reliable evaluation of the other one. This Reichenbach loop is tantamount to a self-referential distortion. The one-way transport velocity, and the relativistic time lag going along with it, constitutes a self-referential distortion of any measurement of distant simultaneity. In order to defuse the circularity inherent in this situation we applied the same strategy as we did in II.5 for coping with self-referential distortions. The pivot of that strategy was to identify an experimental set-up in which the distortion was no longer present or no longer influential. Remember the example of the Wheatstone bridge (cf. II.5.§1). In that example a galvanometer was applied in such a fashion that its distorting internal resistance had no effect on the outcome of the measurement. Slow transport synchrony is conceptually analogous to the Wheatstone bridge. Equation (10) specifies the pure case ($\Delta t = 0$) as transport with velocity zero. Under this condition the distorting relativistic time dilation vanishes and the raw-readings of the transported clock may legitimately be used for synchronizing the distant clock.

The pure case is realized, then, by taking advantage of Equation (11) along with its implementation by an iterative process. Accordingly, just as described in II.5.§4, the pure case is made accessible empirically through *external fixing*. An additional law, namely, Equation (11), is brought to bear in order to get hold of the pure case as implied by Equation (10). What is remarkable though is that both equations can likewise be derived from the basic time dilation formula of Special Relativity (cf. Ellis/Bowman 1967, 128, 130). So they are not really independent of one another. This procedure thus represents a particularly parsimonious type of external fixing. Two different consequences of the same fundamental law are employed. It will be seen in § 6, however, that there is, strictly speaking, an additional hypothesis needed for the identification of the pure case. Anyway, by means of this method one arrives at a definite relation of distant simultaneity which is subsequently used for an evaluation of the distorting factor, i.e., the one-way transport velocity, in all non-ideal cases. And on that basis one can calculate the relativistic time lag from Equation (10) and thus easily correct for non-vanishing transport influences.

Correspondingly, just as in the examples discussed in II.5, the correction circularity is transformed into a two-step linear procedure. First, the pure case is outlined and the magnitude of the distorting effect is determined on its basis. Second, the result obtained is used for performing the corrections in the distorted cases. The only novel conceptual element in the clock transport scenario is that, in contrast to the cases discussed in II.5, the pure case cannot be instantiated. Transport with vanishing transport velocity is somewhat hard to put into experimental practice. Yet the applicability of the correction procedure is rescued by resorting to the strategy of successive approximation. Use of this strategy licenses the inference to the pure case which is itself inaccessible to direct inspection. The conclusion is that Special Relativity provides a basis for overcoming the self-referential circularity underlying the empirical determination of distant simultaneity. And this means that Special Relativity is *complete* with respect to measurements of distant simultaneity.

§ 4. Yet there exists another route to the same destination. Clock transport can also be used in a quite different fashion for the purpose of specifying a preferred relation of distant simultaneity. The crucial element of the method sketched in § 3 was the invocation of the pure case, and what might appear objectionable is that this pure case is not directly realizable experimentally but is only accessible through extrapolation. However, the just-stated conclusion as to the completeness of Special Relativity can also be reached by relying exclusively on actually realizable transport velocities. As I said in II.5.§4, there should exist self-referential distortions which, in virtue of being corrective Reichenbach loops, are amenable to non-instantiative treatment. We will see presently that transport synchronization is a case in point.

At bottom, the relevant procedure is a clock-transport version of the two-way light signal experiment outlined in IV.4.§6. Consider two distant clocks C_A and C_B at relative rest. We equip ourselves with two more clocks C_C and C_D which move inertially relative to this rest frame, one from C_A to C_B and the other from C_B to C_A. In particular, we proceed as follows. At the start, C_C is locally synchronized with C_A and subsequently transported with some as yet unknown velocity to

C_B. There, things are arranged so that at the very instant C_C arrives at C_B, the fourth clock C_D also passes C_B, but in the opposite direction. The fleeting moment of encounter is used for locally synchronizing the outgoing C_D with the incoming C_C. The reason for employing this apparently clumsy two-moved-clocks procedure is to realize a round-trip transport from C_A to C_B and back to C_A without having to accelerate a clock. This avoids complicating things with the General Relativistic time lag of accelerated clocks (cf. IV.3.§2). In the version described, the method makes exclusive use of inertially moved clocks.

It follows from Special Relativity, or more specifically from Equation (10), that upon the return of C_D to C_A the time readings of the two clocks do not coincide. Due to relativistic time dilation, the initial synchrony has been destroyed by the transport, and C_D lags behind C_A by the amount Δt, as given by Equation (10). Now the trick essentially is to infer the transport velocities involved from the value Δt measured locally at C_A.

This can be accomplished as follows. We assume standard synchrony between C_A and C_B in order to be able to speak of the one-way velocities of the transported clocks and to derive the predictions Special Relativity makes about the concomitant dilations. The experiment consists in performing several round-trip clock transports with different back and forth one-way velocities. The *constraint* imposed on the experiment is that the total round-trip time of C_C and C_D, as measured with C_A, is constant. That is, in one case the forward velocity of C_C is small while the return velocity of C_D is large, in another one the ratio is reversed, in still another one the two are moved with equal velocities, and so on. We combine a number of different to and fro transport velocities so that the average round-trip speed is always the same.

Special Relativity predicts, then, that the time lag Δt of C_D upon return to C_A is dependent on the particular transport velocities chosen. This is best understood by considering an example. Let s denote the distance between C_A and C_B, as measured in their common rest frame, and let v_t and v_f designate the to and fro velocity of C_C and C_D, respectively. The velocities are determined using the time readings of C_A. The time dilation of C_D upon its arrival at C_A is obviously the sum of the time lags induced by the outward and the return motion. Equation (10) yields for this quantity:

$$(12) \quad \Delta t = \frac{s}{2c^2}(v_t + v_f)$$

Furthermore, the periods of back and forth transport must add up to the fixed round-trip travel time T. That is:

$$(13) \quad \frac{s}{v_t} + \frac{s}{v_f} = T$$

Consider now two particular cases in order to realize what these equations entail. Suppose we first make a round-trip experiment with the to and fro velocities both being equal to $c/4$. The resulting time lag Δt comes out as $s/4c$. Then pick a forward velocity of $c/2$. In order to satisfy the constraint (13), the return velocity must be equal to $c/6$. These values yield a time lag of $s/3c$ which is larger than the one obtained before.

This observation can be generalized. It has been proved by Salmon, in collaboration with Grünbaum and Janis, that Special Relativity entails the following theorem regarding such differences in retardation. *Minimum retardation theorem*: The back and forth time lag or round-trip retardation assumes a *minimum* value if the outgoing and return velocities are *equal* in magnitude as measured in standard synchrony.[36] This noteworthy result allows for specifying the following synchronization procedure. As a starter we drop the earlier assumption that C_A and C_B are in standard synchrony; it is this synchrony relation that the procedure is supposed to end up with. Then we consider a series of pairs of clocks with identical average round-trip speeds, as measured by C_A, but different yet unknown one-way transport velocities. We record the respective time lags of the various incoming clocks with respect to C_A. Subsequently, we pick the case with minimum time lag (by interpolation if necessary) and conclude that in this case the to and fro standard velocities are equal in magnitude. It is clear, then, that in this case the distant event of swapping clocks is simultaneous to the mid-point between the time of departure from and the time of return to C_A. The distant clock exchange divides the round-trip time T into two equal halves. And this is precisely constitutive of standard synchrony (cf. Salmon 1969, 47–48).

The logic of the argument is thus as follows. Given the laws of Special Relativity, equal back and forth one-way transport velocities, as measured in standard synchrony, result in minimum time dilation. Consequently, minimum time dilation indicates equal standard transport velocities. This reveals that cases of equal one-way transport velocities can empirically be identified prior to measuring one-way transport velocities in general. And this allows for stipulating distant simultaneity without the need to evaluate any such velocity beforehand. After distant simultaneity has been established in this way, arbitrary one-way transport velocities can be measured.

§ 5. The systematic significance of this method is as follows. The minimum retardation theorem is used for identifying equal standard velocities by *exclusive* reference to transported clocks; clocks, moreover, whose transport velocities can actually be realized (or instantiated). And this means that clock transport is sufficient for singling out in an operationally feasible fashion distant events exhibiting the relation of standard synchrony. That problem was, it will be recalled, that the identification of any such events appeared to be precluded since employing uncorrected clock readings leads to inconsistent results, on the one hand, and applying a correction procedure is vitiated by a Reichenbach loop, on the other (see § 1). Salmon's round-trip retardation method can thus be reconstructed as the demonstration that this corrective Reichenbach loop between the evaluation of the one-way velocities of clock transport and the determination of distant simultaneity is actually not vicious.

Of course, one could have proceeded alternatively by first defining distant simultaneity by means of light signals, using the standard convention of isotropic light propagation, and by measuring afterward the one-way transport velocities. This alternative comes down to breaking up thc Reichenbach loop by establishing one of the relevant quantities, namely, distant simultaneity, relying on an additional procedure independent of clock transport. It comes down, that is, to external fixing (cf. II.4.§4). The loop is dissolved by resorting to a further hypothesis that removes the reciprocal dependence between the two quantities in question. Salmon's round-trip retardation method makes it clear, by contrast, that as regards clock transport synchronization, recourse to an auxiliary procedure of the kind is unnecessary. Put conversely, it is indeed possible to synchronize distant clocks in the standard manner without having to rely on any procedure other than clock transport. The Reichenbach loop is dissolved in a self-reliant and thus more complete fashion. We can use clock transport for singling out empirically the relation of standard synchrony among various locations within an inertial frame. The distortions inherent in clock transport can be rectified using only the readings of transported, i.e., distorted, clocks.

But how was this result obtained precisely? The only known alternative to external fixing is non-instantiative testing, and in fact the central step in Salmon's procedure is non-instantiative in kind. This may sound strange for we end up with definite instantiations of distant simultaneity and one-way transport velocities which obviously doesn't fit the non-instantiative scheme. Still, the core element of the procedure, namely, the minimum retardation theorem, is a special consequence of Equation (10), not an instantiation of it. Equation (10) relates the one-way time lag to the one-way transport velocity and neither quantity figures in the retardation theorem. To be sure, the latter furnishes an instantiation of the *equality* of one-way velocities, but this is still compatible with a host of actual values of these velocities. It is this consequence relation that makes up the non-instantiative step.

What is peculiar, however, is that this consequence relation cannot definitely be instantiated itself. For it provides the *only* access to equal one-way velocities; the latter cannot be instantiated independently. The minimum retardation theorem thus represents a Sneed bottleneck. This is why we are not dealing with a case of non-instantiative testing proper; the consequence relation bears no risk of empirical failure (cf. II.3.§5). And I don't see a way to supply it with empirical content in a non-instantiative fashion; we may of course resort to a convention about light propagation (as just described) and thus to external fixing. Anyway, since our primary concern is not to check the truth of Equation (10), but rather to ground an instantiation procedure on it, this test problem isn't particularly worrisome.

The final step is to bring to bear Equation (13) expressing a kind of extensivity constraint for temporal intervals: The temporal intervals of to and fro transport add up to the total time elapsed. The application of this constraint to equal one-way velocities yields the conclusion that the distant clock exchange is simultaneous to $T/2$, as measured by C_A. This constitutes an instantiation of a distant simultaneity relation which can subsequently be used for evaluating arbitrary one-way velocities.

There is one last feature of this procedure worth mentioning. As I argued in § 1, the Reichenbach loop inherent in transport synchronization can indeed be interpreted as a self-referential distortion. The clock transport produces a time lag that distorts measurements of distant simultaneity in a self-referential way. The procedure just outlined thus establishes the possibility, already anticipated in II.5.§4, of rectifying a self-referential distortion without having to base the correction process on the undistorted case. After all, whereas the slow transport method of dissolving the self-referential circularity only succeeded by resorting to an experiment in which no correction is needed, the round-trip transport method does not invoke the pure case of vanishing time dilation.

§ 6. The final issue to be addressed in this chapter is the bearing of these two methods on the problem of the conventionality of distant simultaneity. To anticipate the answer right at the start: It has none. The point of these two methods merely is that they direct us on how to put the *decision* to use standard synchrony into experimental practice without getting stuck in a circularity. Apart from that, nothing is achieved by these methods, and the reason is that they leave ample room for gerrymandering. Let's have a closer look at how we could still stick to the view that some non-standard ε-value represents an acceptable synchrony relation.

As regards slow transport synchronization, the chief reason for presuming that this method gives empirical privilege to standard synchrony is that Special Relativity in virtue of Equation (10), on which the method is based, seems to require that the time lag between the stationary and the infinitely slowly transported clock vanishes. And since this result implies standard synchrony, the latter relation appears to be a direct consequence of that theory. Standard synchrony should thus be just as factual as Special Relativity as a whole.

The catch in this argument is, however, that Equation (10) is already infected with standard synchrony. From a formulation of Special Relativity that is non-committal concerning that issue it only follows that in the limiting case of infinitely slow transport the time lag between stationary and transported clock approaches the value $[s(2\varepsilon - 1)]/c$ (which obviously reduces to zero if ε is chosen to be $\frac{1}{2}$) (cf. Winnie 1970, 226). Furthermore, the Lorentz transformations which in their usual formulation entail $\varepsilon = \frac{1}{2}$ can be rewritten so as to allow for arbitrary choices of ε (within the limits given by Equation (9) (cf. III.4.§5)) (cf. Winnie 1970, 231–237). These and similar results support the conclusion that given the truth of the empirical consequences of Special Relativity, or given Special Relativity without one-way velocity convention, we are free to cling to our favorite simultaneity relation.

Accordingly, a disagreement between the readings of an infinitely slowly transported clock (which coincide with standard synchrony) and those of a non-standardly synchronized stationary clock in no way automatically disqualifies the latter readings. One may consistently adopt the point of view that the former clock readings fail to establish a relation of synchrony without thereby running into conflict with the convention-free formulation of Special Relativity. This amounts to arguing that the readings of the slowly transported clock have to be corrected in order to be in agreement with those of its non-standardly synchronized stationary counterpart. It is the moved clock that has to give in, not the stationary one. In sum, if one decides to regard slow clock transport as a synchronization procedure as it stands, standard synchrony follows. But the obeservable facts do not oblige us to regard it that way (cf. Salmon 1969, 51; Grünbaum 1973, 697; see also Winnie 1970, 226–228). Accordingly, the additional hypothesis underlying slow clock transport synchronization, as announced in § 2, is actually $\varepsilon = \frac{1}{2}$.

Systematically speaking, it is the nature of the pure case that is under dispute here. According to standard synchrony, it is the slowly transported clock that represents the undistorted case (since time dilation vanishes), whereas a deviant light signal synchronization involves unisotropic light propagation and is thus thwarted by an unrecognized perturbation. On the non-standard view, by contrast, the judgment as to which constitutes the pure case and which the distorted one is reversed. But if there is no agreement about what is to pass as the pure case, the strategy of overcoming a self-referential circularity through the realization of the pure case is clearly of no avail. As a solution to the problem of the conventionality of distant simultaneity, slow clock transport caves in altogether.

Regarding the same problem, round-trip transport scores equally bad. The reason is that the minimum retardation theorem is in no way suited to give standard synchrony primacy over its alternatives. Let's explore how we could still get away with a non-standard choice of ε. The round-trip time lag to which the theorem refers is measurable with a single clock and is thus synchrony independent. Consequently, on choosing a non-standard ε-value, the same pair of moved clocks is picked. But this pair is no longer distinguished by equal back and forth transport velocities. All one-way transport velocities are affected by our choice of ε; in particular, the outgoing and return ε-velocities are different in the case under consideration.[37] Consequently, it would be illicit, on that non-standard approach, to ground the synchronization of the distant clock C_B on this case. Rather, we are directed to appeal to the case of equal ε-velocities for this purpose. It is the latter case of equal outward and backward ε-velocities that constitutes the legitimate basis for a synchronization procedure. But since this case does not coincide with the original scenario of equal standard velocities, the synchrony relation we will end up with is non-standard.

This means that the minimum retardation theorem is ill-suited for supporting a certain choice of ε to the exclusion of others (as Salmon is at pains to stress (cf. Salmon 1969, 50)). The validity of the theorem is restricted to standard velocities; i.e., it is not true that the round-trip retardation assumes a minimum value if the clocks are transported with equal ε-velocities. This reveals that the derivation of this theorem invokes a further constraint, apart from the innocuous constraint expressed by Equation (13) (see § 4); it invokes, namely, the concealed appeal to $\varepsilon = \frac{1}{2}$. The decision to adopt the theorem as a foundation of distant clock synchronization thus implicitly contains the commitment to standard synchrony. To use the theorem as an argument in favor of the latter thus amounts to going full circle.

§ 7. In fact, it can be proved in full generality that Special Relativity prohibits specification of any method that gives empirical primacy to standard synchrony. Special Relativity is reducible to General Relativity. The former can be derived from the latter under the condition that the Riemann curvature tensor vanishes, i.e., that space-time is flat. It is possible to introduce a global inertial frame into such a space-time and in that frame the metric will come out in its familiar Minkowskian form.

Since Special Relativity is part of General Relativity in this sense, it appears that the former, just like the latter, can be given a generally covariant formulation (cf. IV.3.§1). That is, although it is possible to describe all physical events using a global inertial frame, it is by no means mandatory to do so. Rather, one may equally well resort to accelerated frames of reference and choose curved spatial coordinate lines and non-standardly coordinatized time axes. Now, the essential point is that picking deviant definitions of distant simultaneity is tantamount to adopting a peculiar coordinate system of the kind. The possibility of choosing $\varepsilon \neq \frac{1}{2}$ is thus an immediate consequence of the fact that Special Relativity can be formulated without recourse to any distinguished state of motion (cf. Havas 1987, 438–439; see also Friedman 1977, 418–421; Friedman 1983, 175).

It has been argued that the conventionality based on general covariance is rendered trivial by an important distinction standard synchrony receives from a different source. In flat space-time the relation of standard synchrony, and *only* this relation, can be defined from the fundamental relation of causal connectibility, i.e., connectibility through light signals. More specifically, recourse to the light cone structure is sufficient for grounding standard synchrony, in contradistinction to any other ε-choice (cf. Malament 1977, 297–299; Friedman 1983, 310–311).

Even so, the conventionalist need not feel obliged to eat crow. First, the methods employed in the actual construction of the Minkowski metric (along with its built-in standard synchrony) are essentially global in character and do not admit transformation into a local, operationally feasible procedure (cf. Winnie 1977, 187; see also Ehlers 1973a, 81). Second, these methods work properly only in flat space-time. In the more comprehensive case of general relativistic space-times, the metric ceases to be causally definable anyway (cf. Winnie 1977, 195–197; Sklar 1977, 249). Third, it is true that every non-standard choice of ε requires us to bring to bear extra ingredients, not definable from the causal structure, so as to identify the coordinate system we prefer to pick. But what does this feature really imply? Neither does it constitute an experimental fact nor is it given by divine decree that simultaneity relations have to based exclusively on causal relations. It is fine that we can do so, but ability does not imply obligation.

In fact, underlying the felt obligation to restrict oneself to causal means, if possible, is the methodological commitment to giving a theory the most simple and parsimonious formulation that the facts admit. This comes out clearly in Michael Friedman's argument that every non-standard choice of ε is discredited, empirical equivalence notwithstanding, by the fact that it brings an otiose machinery in its train, a machinery whose rods and gears run idle (cf. Friedman 1983, 312). The observation underlying this argument is quite true. Non-standard ε-choices are vastly inferior as to this methodological virtue, and they are not distinguished by any other attractive features that could possibly outstrip this defect.

But pragmatic criteria such as parsimony and simplicity are simply not at issue, as has been stressed time and again by conventionalists.[38] And this is all the more important if the credentials advanced in favor of standard synchrony boil down to the temporal analog to preferring Cartesian coordinate systems over oblique ones with a fixed angle.[39] The conclusion accordingly is that the conventionality of distant simultaneity is left unscathed by these arguments. It is a different question, however, whether this conventionality is appropriately to be regarded as an interesting or deep-searching peculiarity or whether it derives from a spurious Reichenbach loop. For given the methodological inferiority of all non-standard synchrony relations it seems straightforward simply to amputate that loop altogether – in line with analogous operations performed in II.4.§7. I will resume the discussion of this issue in VI.1.§6.

6. CONCLUSION

Time has come to take stock again. In the foregoing discussion two levels of consideration were addressed. On the first level, influences that transcend the confines of the standard formulations of the pertinent theories were at issue. We were concerned with the import of universal forces on the determination of the metric and the geodesic structure and likewise with the import of unisotropic light propagation *in vacuo* on the relation of distant simultaneity. In the standard formulations of the corresponding theories (i.e., General and Special Relativity, respectively) these influences have already been removed; these standard versions proceed on the assumption that universal forces vanish and that the one-way velocity of light *in vacuo* is neither direction nor position dependent. So, what we did was to explore the relation between these standard formulations and alternative renderings, the latter being likewise able to accommodate the relevant facts albeit in a methodologically inferior fashion.

The point is, then, that the relations between quantities figuring in these non-standard accounts instantiate the pattern of a Reichenbach loop. Each pair of quantities in question is given by only one supposed law with the result that the evaluations of these quantities are reciprocally dependent on one another. This kind of relation obtains between the non-standard metric and a universal force (cf. IV.4.§3), the non-standard affine connection and a universal field (cf. IV.4.§5), and the non-standard simultaneity relation and the one-way velocity of light (cf. IV.4.§6). By virtue of this reciprocal dependence, these cases qualify, on the face of it, as instances of strong completeness on the hypothesis level. And it became clear that this peculiarity indeed brings serious test problems in its train. The respective Reichenbach loop induces a vicious circularity between the evaluations of the quantities involved with the result that no particular evaluation can be given primacy over an alternative one on empirical grounds. We are thus presented with non-Duhemian test-problems. And it turned out that in the cases presently under consideration these non-Duhemian test-problems, along with the Reichenbach loops underlying them, could not be dissolved; the circularity is here to stay (cf. IV.4.§4, §7; IV.5.§6). We have thus come across an instance of *failed* completeness. Definite instantiation of one of the quantities in question demands that the underdetermination problem be solved. But the pertinent basic hypotheses do not provide a means suited for this purpose. This is how the conventionality claims are to be reconstructed in terms of completeness. The conventionality of physical geometry and of distant simultaneity are manifestations of attempts to establish strong completeness on the hypothesis level, attempts that come to grief. Reciprocal dependence produces untestability of the underlying hypotheses. As a result of these truly vicious Reichenbach loops, we are allowed to give free rein to our whims and predilections.

On the second level, the discussion was concerned with those influences that are recognized within the framework of the standard formulations of the respective theories, or for that matter, the Special Theory. It was the feasibility of clock synchronization by means of clock transport relying on Special Relativity that was at issue. The influence threatening the practicability of this method is relativistic time dilation. Time dilation affects the rate of a transported clock and thus constitutes a distortion that has to be reckoned with in establishing transport synchronization. However, rectifying this distortion requires reference to the one-way transport velocities which in turn can only be determined by having recourse to distant simultaneity – the very quantity we are after. So, one more Reichenbach loop emerges. This one grows out of the reciprocal dependence between the measurement of one-way transport velocities and the corresponding relation of distant simultaneity. On the face of it, we are again faced with a case of strong completeness on the hypothesis level that ushers in a Reichenbach loop. So we may suspect that, just as before, there is no way to distinguish empirically any particular value of the relevant quantities. Apparently, clock transport is unfit for providing a method for establishing a definite relation of distant simultaneity.

It turns out on closer scrutiny, however, that this suspicion is unjustified. In fact, there are even two methods which can be successfully applied for this purpose. In the framework of the first one, the relevant Reichenbach loop is construed as a self-referential distortion, and the necessary correction is performed unambiguously by resorting to a realization, albeit by successive approximation, of the pure case. In this way the general method for coping with self-referential distortions, as elaborated in II.5, is brought to bear on the apparent circularity inherent in the clock transport scenario. The Reichenbach loop is defused through instantiation of the pure case which is in turn identified by external fixing (cf. IV.5.§3).

The second method, by contrast, does not proceed on the basis of the pure case but rather exploits an empirically accessible consequence of the basic relation between transport velocity and time dilation, which – due to the Reichenbach loop involved – is not empirically testable itself (cf. IV.5.§1). This method treats the Reichenbach loop in an essentially non-instantiative fashion, although it invokes an additional constraint so as to yield a definite evaluation of distant simultaneity (cf. IV.5.§5).

This consideration suggests that there exists a remarkable difference between the first and the second type of cases under discussion. Despite the fact that the logical structure of the problems underlying both types coincides in that it is equally represented by a Reichenbach loop, the conclusions to be drawn from considering each type are discordant. In the first type of cases the corrections cannot be performed unambiguously so that completeness is missed. In the second type, by contrast, they can be performed so that completeness is accomplished. So the question suggests itself naturally: What is the origin of this noteworthy difference?

The answer is that in order to make the two methods function properly some prerequisites have to be satisfied. And these prerequisites are actually satisfied in the second type – in contradistinction to the first one. It is clear that the pure case strategy for handling self-referential distortions demands at the very least that all parties to the discussion agree about what actually represents the pure case. This agreement is present in the second type since all relevant parties concur that in infinitely slow transport the distorting influence (namely, time dilation) reduces to zero. In the more fundamental debate about the conventionality of distant simultaneity, this agreement can no longer be presupposed. By contrast, it is the very point of the whole debate that disagreement on that issue is possible while doing equal justice to the empirical record. It is perfectly possible to regard the case of direction-dependent one-way signal velocities as the pure one, and to claim, accordingly, that the infinitely slowly transported clock suffers from a time dilation (or "time contraction", as the case may be) (cf. IV.5.§6). Analogously, a universal force scenario differs from its geometrized counterpart by considering cases as distorted which the latter regards as pure, and conversely (cf. IV.4.§4). It is not astounding that given such differences in assuming what is unperturbed and what is not, every method that relies essentially on the identification of the pure case is doomed to failure. It can obviously contribute nothing to settling this debate.

The same goes analogously for the non-instantiative strategy. Its application crucially rests on the possibility of deriving a consequence relation from a hypothesis that – due to a Reichenbach loop – cannot definitely be instantiated itself. In the round-trip clock transport scenario it is the minimum retardation theorem that constitutes the relevant consequence relation. The trouble is, however, that this theorem is not accepted by non-standard theoreticians since its validity is confined to standard synchrony. The derivation if this theorem is critically dependent on assuming standard synchrony in advance. This can be expressed thus that in this derivation $\varepsilon = \frac{1}{2}$ is appealed to as a constraint (cf. IV.5.§6). And since this assumption is precisely what non-standard theoreticians reject, they will hardly feel defeated by this argument. If there is no agreement about what qualifies as a soundly derived consequence relation, non-instantiation finds no hook to lock into.

So in sum it is disagreement about what constitutes the undistorted case and about what counts as an acceptable constraint that vitiates the applicability of our two strategies to defuse Reichenbach loops. It is this feature that underlies the conventionality of physical geometry in general and of distant simultaneity in particular. Conversely, if this disagreement is removed Special Relativity suffices for the specification of operational procedures suitable for identifying distant events in standard synchrony – in spite of the Reichenbach loop involved in these procedures. Special Relativity is actually able to cope with distorted cases in its domain of application. This is why Special Relativity is complete with respect to the measurement of distant simultaneity.

But this is obviously not the whole story. For we haven't yet talked at all about the functioning of clocks. Completeness demands, it will be recalled, that we are in a position, trusting the theory's guidance, to pass over from that theory's central concepts to the associated pointer readings. But it is clear that we have not reached this goal yet. A quartz clock isn't governed by the laws of Special Relativity. Accordingly, the above assessement as to the completeness of this theory has to be qualified. In fact, it is precisely this insufficiency of Special Relativity to deal with the internal operation of clocks that Einstein complained about in the passage that first circumscribed the notion of completeness (cf. II.1.§2). It will turn out, however, that this complaint is no longer justified. For since that time, clocks have been constructed that rely (almost) exclusively on processes which Special Relativity actually covers. I will discuss such *light clocks* in the more comprehensive context of General Relativity that is to be addressed in the following chapter.

CHAPTER V

THE COMPLETENESS OF GENERAL RELATIVITY THEORY

In the present chapter I discuss the potential completeness of General Relativity in its standard version. In contradistinction to the preceding chapter, I do not consider alternative formulations of the theory but rather assume throughout that universal forces vanish and that $\varepsilon = \frac{1}{2}$. Given these features, the question is, how does General Relativity fare regarding the specification and analysis of procedures suitable for translating its abstract geometric machinery into concrete experimental data?

There are two types of geometric notions involved here; first, the geodesic trajectories free particles are supposed to pursue (cf. IV.3.§4) and, second, the length and duration relations realized. So, the geodesic and the metric structures of space-time are at issue. These structures constitute the evidential basis of physical geometry which the field equations of General Relativity are intended to accommodate. The challenge is, accordingly, to translate the – to all appearances – heavily theory-laden concepts of geodesic path and metric interval into an empirically feasible measuring procedure relying as much as possible on the theoretical resources of General Relativity. Both issues of identifying geodesics and determining metric intervals will turn out to be closely intertwined in fact.[1]

1. MEASUREMENTS BASED ON PARTICLE PATHS AND LIGHT RAYS, OR: THE ART OF MINIMIZING LABORATORY EXPENSES

§ 1. Our concern in the present chapter is with measurements. Measurements may be performed under *ideal* or *distorted* conditions. Regarding General Relativity, there are two types of distortions left after Reichenbach's universal forces and non-standard ε-values have been dropped. The first one concerns "geometric" distortions, i.e., effects that are in principle part of General Relativity's domain of application. Geometric distortions are generated by properties of a body that are sensitive to space-time curvature. A body's extension or its rotational state are examples in point. These properties interact with a possibly present curvature with the result that the corresponding body is deformed or driven off its geodesic course.

Distortions of another type are substance-specific in kind; they vary with the particular material under consideration. A case in point is electric charge. This type may correspondingly be characterized as being constituted by *differential distortions* (cf. IV.4.§1). In the present section, I address the attachment of indicator phenomena to the assumed geometric states both under ideal circumstances and in the presence of geometric distortions. In the following section, I turn to the problem of establishing such empirical indicators under differentially distorted conditions.

For the discussion to follow, an ingenious approach, developed by Jürgen Ehlers, Felix Pirani and Alfred Schild in 1972, to "grounding" the empirical basis of General Relativity in an inductive fashion on "simple experiences" is of great importance. The pertinent simple experiences are particle trajectories and the paths of light rays. The space-time structure as assumed by General Relativity is intended to be established by an "inductive ascent" that starts from allegedly non-theory-laden facts of this sort. It is recognized immediately that the spirit of such an attempt to give General Relativity an "operational foundation" is in striking conflict with the ideas underlying completeness (cf. V.4.§1, §3–4). Still, it will turn out that the technical insights gained by Ehlers–Pirani–Schild can be used (or abused, judging from their point of view) for assessing the completeness of General Relativity. The point of contact is that an operational approach attempts to dispose as much as possible of theoretical ingredients in carrying out and interpreting experiments. It thus derivatively attempts also to dispose of external observation theories (cf. V.3.§1). It is this latter goal in which an operational approach and the completeness project meet. We will come to realize that the operational emphasis on measurements that can be performed and analyzed on a fairly restricted conceptual basis serves to prepare a fertile ground for examining the completeness of General Relativity Theory.

§ 2. As just indicated, Ehlers–Pirani–Schild set themselves the task of establishing the evidential basis of General Relativity, and in particular the metric structure of space-time, without invoking that theory's conceptual apparatus. In order to fulfill this task, they proceed roughly as follows. The first thing to do is to put ourselves in a position to keep track of trajectories. This means, we need a coordinate system. On the other hand, we are not allowed to employ rigid rods and metric clocks for this purpose since such instruments are certainly not directly given by experience. They are supposed to be provided by the whole construction and can therefore not be invoked beforehand.

We restrict ourselves to the motion of "test particles," i.e., particles in the limit of vanishing mass and vanishing extension. We begin by parametrizing the world-line of an arbitrarily moved test particle using a "non-metric" clock; a clock, that is, running continually and smoothly. The readings of such a clock are not required to mirror metric intervals (in particular proper time); rather, it is merely demanded that different events on the world-line at hand are attached different clock readings, and that adjacent events are ascribed neighboring temporal coordinate values. Next we pick an event to be coordinatized and choose two nearby particle world-lines parametrized in the way described. We send off a light signal from each of these world-lines and record its emission time and the time of its return after its supposed reflection at the event under consideration. The four temporal coordinate values obtained in this fashion yield the basis for a local coordinate system; and various such systems can be conjoined to form a so-called atlas. By means of these "radar coordinates" every physical event can be assigned a spatiotemporal location.

The second step of the construction takes advantage of the particulars of light propagation. The light rays connecting a given event with events nearby form a cone structure if represented spatiotemporally. The light-cone structure thus obtained allows for distinguishing between spacelike and timelike four-directions, and it can be shown that the metric is determined by this structure up to a positive function of the four-positions. That is, the examination of light propagation considerably reduces the set of empirically admissible metrics. Moreover, light rays turn out to be null geodesics with respect to this class of admissible metrics.

In the third step the motion of freely falling particles is taken into consideration. This motion determines the straightest spatiotemporal lines possible, i.e., the timelike geodesics of the respective space-time. And the challenge is to identify such geodesics non-metrically. In fact, the challenge can be met. A geodesic can be characterized by two conditions. It is, first, *unique* in the sense that the entire trajectory is unambiguously determined by the four-direction of the trajectory at one point on it. This condition is obvious in the special case of a straight line in flat space-time and the just-given clause is a natural generalization of that special case to situations with non-vanishing curvature. Accordingly, for a given event, a given initial velocity and a given spatial direction there is precisely one world-line representing free fall motion.[2]

The first condition, however, is not alone satisfied by geodesic motions but also by particle motions under the influence of a constant electromagnetic field – provided that the particle's specific charge is fixed. So the second condition must carry the bulk of the weight. This condition invokes a *local symmetry* of geodesics that can be characterized roughly and intuitively as follows. Consider a collection of path elements or curves that are all passing through a given event. Disregard, moreover, the special parametrizations of these path elements or, to put it the other way around, construe every path element as an equivalence class of differently parametrized curves. Then introduce a "dilation operation" that radially stretches or contracts the path elements away from or toward the event under consideration. If this operation leaves the paths invariant, i.e., if it amounts to a map of each path onto itself, these paths are geodesics. After all, it's intuitively clear that the only paths that are left unchanged by such a dilation procedure are radial straight lines, and it has been shown by Ehlers and Köhler that the same feature, if interpreted locally, is likewise characteristic of their analogs in non-flat geometries, namely, geodesics. A micro-symmetry of this sort is sufficient for singling out geodesics in every space-time manifold. This means that the motions of freely falling particles provide a basis for establishing the geodesic structure of space-time.

It has been demonstrated by Heilig and Pfister that this third step of identifying geodesics can also be carried out in an alternative fashion. Their approach heavily relies on Desargues' theorem of projective geometry. The underlying Desargues figure can be realized using ten straight lines and ten relevant points of intersection between these lines. In flat Minkowski space-time a spatiotemporal version of this figure can be constructed by means of straight particle paths. The reverse holds as well: If the Desargues figure can successfully be constructed from particle trajectories, then these trajectories are straight and the corresponding particles are consequently inertially moved. The salient aspect is that this criterion is indeed generalizable to the pseudo-Riemannian space-time of General Relativity. It can be shown, namely, that a local version of Desargues' theorem approximately holds under these circumstances and that the condition of local Desargues constructibility uniquely determines freely falling particles. That is, if the Desargues figure can successfully be built up locally using particle trajectories, then three of the trajectories involved are demonstrably geodesic.

The fourth and final step of the construction is to tie together the results of the second and the third step, i.e., to connect the light cone structure to the geodesic structure obtained from free fall motion. This is accomplished by way of a compatibility postulate to the effect that the null geodesics as provided by the light cone structure are likewise null geodesics with respect to the geodesic structure resulting from particle motions. This means that the trajectories of freely falling particles always remain within the corresponding light cone although they may come arbitrarily close to it. This expresses the condition that particles cannot travel faster than light although the speed difference may become arbitrarily small. In this way the space-time structure of General Relativity can be derived from considering the features of light propagation and free fall motion.[3]

§ 3. So far, so good; but what is the impact of this approach on our endeavor to examine the completeness of General Relativity? The first thing to be noted is that the approach supplies us with a working procedure (or rather two of them) to identify freely falling test particles. I mentioned in § 1 that geometric distortions lead to the deviation of a particle from its geodesic trajectory even if no external forces are present. Extended or rotating bodies do not move along geodesics in a gravitational field. Nor do bodies endowed with a gravitational quadrupole moment. The Ehlers–Pirani–Schild method (along with the supplements sketched) demonstrates that such distorted motions can effectively be identified and consequently be removed from consideration. This method thus enables us empirically to get hold of the relevant pure states of motion without relying on the field equations of General Relativity (and even, according to their contention, with a minimal invocation of other observation theories).

Furthermore, there is one aspect of the Ehlers–Pirani–Schild procedure for singling out timelike geodesics or inertial motions that deserves emphasis. The traditional approach to the law of inertia amounts to an existence claim regarding privileged coordinate systems: There exists a preferred class of coordinate systems, namely, inertial frames of reference, in which the acceleration values of inertially moved particles vanish (cf. IV.3.§1). By contrast, the methods described in § 2 identify geodesics through their internal or geometric properties. Inertial motions are characterized without distinguishing in advance any preferred frame of reference.

The conclusion is that we can indeed identify free fall motion without prior recourse to a privileged class of coordinate systems and without relying on specific accounts of possible geometric distortions. So, for the time being we are done with geodesics (although we will have to reconsider the matter in sec. V.2). What remains to be discussed is whether the Ehlers–Pirani–Schild procedure provides a basis for performing actual geometric measurements. What we need is an instrument suitable for detecting the prevailing metric relations.

§ 4. In order to comply with the spirit of the Ehlers–Pirani–Schild-procedure such an instrument has to make exclusive use of test particle paths and light rays. The central metric quantity in question is proper time, and thus our task comes down to measuring this quantity by recourse to a *light clock*. The historically first clock of the kind is the so-called *Marzke–Wheeler clock*, developed prior to (and thus independent of) the formulation of the Ehlers–Pirani–Schild program. In essence, a light clock consists of a light pulse reflected back and forth between two timelike world-lines of unchanged four-distance. Think of two mirrors moved in the same spatial direction with the same speed. Equal intervals of proper time are then indicated by the successive return times of the reflected light pulse to one of the mirrors.

This sounds easy enough, but the challenge is to specify two world-lines of constant distance without having rigid rods or other length standards at one's disposal. Remember, the metric is supposed to be provided by the clock; so the metric must not enter the clock construction. The Marzke–Wheeler set-up presupposes the *flat* Minkowski space-time of Special Relativity. In the absence of space-time curvature, the timelike geodesics of freely falling test particles and the null geodesics of light rays all become straight lines. The point of the Marzke–Wheeler clock is to specify a procedure for constructing a *parallel* to a given straight line using only theorems of ordinary projective geometry. This means, given straight particle trajectories and light rays, a parallel to a particular straight particle trajectory can experimentally be determined without resorting to any distance measurement (cf. Marzke/Wheeler 1964, 50–52).

With two parallel straight lines at hand it is an easy matter to realize a light clock in the way just described. At bottom, a light clock is represented by a photon bouncing between two parallel particle trajectories. Moreover, spatial distances are now accessible as well; namely, by employing the radar technique. We send off an additional light pulse to the object whose distance we wish to determine and record, using our light clock, the time elapsed between emission and return after reflection. This procedure furnishes a spatial length standard. The method is essentially based on the idea of letting the end-points of a freely falling rigid rod be represented by the two test particle world-lines constituting the Marzke–Wheeler clock. In sum, the Marzke–Wheeler clock enables us to determine empirically metric space-time intervals by relying exclusively on free particle motions and light rays.

At first sight at least, for there is a difficulty involved that deserves closer scrutiny. I mentioned already that the construction only succeeds in flat space-time. But the space-time of General Relativity is curved rather than flat. Curvature thus represents a distortion that spoils the applicability of the Marzke–Wheeler clock. Tidal forces destroy the parallelism of the two world-lines so that the reflected light pulses will increasingly be out of step. Is it possible to cope with this problem? Marzke–Wheeler claim that it is and argue as follows. First, the flatness condition is physically innocuous in the case of a small clock operated in the local inertial frame. That is, the clock works properly if the relevant particles are freely falling, their distance is small and the time intervals to be recorded are short. Second, for transporting the standard interval over longer spatiotemporal distances one proceeds by dividing up the route of transportation into smaller stretches; i.e., stretches small enough to be covered with sufficient accuracy by the corresponding local inertial frame. The clock is used only within a single stretch; that is, within a small spatiotemporal area. And the transport is effected by the successive application of several local clocks.

In practice this means that we use a parallel, as provided by the Marzke–Wheeler procedure, only during a short period of time and then start anew and construct another parallel. Geometrically speaking, this amounts to approximately realizing the parallel to a curved line by a sequence of straight lines each of which is locally tangent to the curve. In this way, it is claimed, the time standard can reliably be transported along a timelike geodesic in curved space-time. Accordingly, the Marzke–Wheeler clock can be employed for comparing distant intervals under these circumstances. It can be employed, that is, for establishing and transporting metric standards. This implies that the clock represents a procedure based on particle trajectories and light rays alone and suitable for a global empirical determination of the prevailing metric relations (cf. Marzke/Wheeler 1964, 53–58). Since no costly instruments such as atomic clocks are needed, this procedure is a real bargain.

§ 5. As already mentioned, the Marzke–Wheeler clock was designed independently of the program of Ehlers–Pirani–Schild, who later drew on the clock in order to complete their attempt to ground the measurement of metric relations on simple experiences (cf. Ehlers/Pirani/Schild 1972, 65; Ehlers 1973b, 34). In the present subsection I examine if and how the clock can live up to these expectations. The point of the discussion is that we encounter once more a pattern familiar from the preceding considerations of completeness problems, namely, a self-referential distortion circularity.

As indicated in § 4, flatness is essential for the proper operation of the Marzke–Wheeler clock. Tidal forces produce a geodesic deviation of the two particle trajectories (cf. IV.3.§7), and the resulting deflection obviously vitiates the correct functioning of the clock. If, for instance, a Marzke–Wheeler clock with a separation of 10 cm between the particle trajectories is operated in a low-altitude orbit around the earth, then this clock is wrong by one light pulse tick after only 0.1 seconds (cf. Ohanian 1977, 909). The measured time interval is thus affected by the intensity of the tidal forces. This entails in particular that the outcome of a transport of the time interval is altered by the tidal forces present along the path of transportation. This effect thus leads to a *route dependence* of transported intervals. Depending on the route we choose for transporting our metric standard, we may end up with different results. And what makes this feature particularly unwholesome is that the Marzke–Wheeler construction does not break down in a tidal field; rather, the results it yields are mistaken. This means that we cannot simply try to set up the clock and see if the construction succeeds. We must make sure in advance that the preconditions for its proper operation are satisfied. So we are compelled to take curvature into account.

In principle, there are two options for doing this. Either we can correct the distorting impact of curvature on the clock readings or we can try to avoid its occurrence in the first place. Let's begin with the first option. In fact, Marzke–Wheeler's proposal to divide the route of transportation into small stretches (see § 4) outlines precisely such a correction procedure. A parallel is replaced by a different one if geodesic deviation has become too large. But in order to apply this correction procedure adequately, we have to know when parallel substitution is called for. We have to know, that is, when geodesic deviation has taken on a magnitude that demands switching to another parallel. This in turn requires that we can assess somehow the strength of the distorting effect. In other words, we must be able to obtain an empirical estimate of the prevailing curvature independent of the results provided by the clock. In particular, it is of no avail simply to measure the relative change in the separation of the two world-lines, since such measurements can be performed reliably only after metric standards are at hand.[4]

However, there exists a method for measuring curvature by recourse to freely falling particles and light rays, namely, the so-called geodesic triangle. The triangle is spread out by two freely falling particles ejected from a common spatiotemporal origin with different initial speeds, for one, and a light pulse sent off from one of the particles to its counterpart, for another. A sufficient number of such triangles allows for the determination of the full Riemann curvature tensor in the region under scrutiny. Unfortunately, however, the actual calculation of the curvature requires measured values for the proper time elapsed between the ejection of the particles and the emission and recording, respectively, of the light pulse (cf. Ehlers 1973b, 51–54). Thus, this method cannot possibly be used for making sure that the conditions of application for our metric clock, designed to furnish proper time in the first place, are met. To my knowledge, every method for measuring curvature likewise requires some metric input.

This result can be put as follows. The influence of curvature on measurements of proper time can only be corrected if proper time has already been measured reliably. The correction procedure can only be carried out if the quantity whose measurement requires correction has already been determined in a trustworthy fashion. This peculiarity obviously signals that curvature is a *self-referential distortion* of proper time measurements (cf. II.5.§3). Underlying this self-referential relation is a *Reichenbach loop* between the empirical determination of curvature and of proper time intervals. We are thus caught in a corrective circle.

§ 6. The principal option for getting out of a self-referential circularity is represented by the realization of the pure or undistorted case. Is there any chance of taking advantage of this option regarding the problem under consideration? In general, geodesic deviation diminishes with shrinking separation of the corresponding geodesics (cf. IV.3.§7). Moreover, its impact is less significant the shorter it exerts its perturbing influence. It should thus be possible to neutralize its effect on the clock's running by making the clock small enough and by operating it during sufficiently short periods of time. By decreasing the clock's spatiotemporal extension we should be able to approach the pure and undistorted case.[5]

This is indeed true, although the actual procedure that is to be invoked for that purpose exhibits a somewhat surprising and unexpected feature. Judging from an analysis pursued by Ohanian on the basis of the Newtonian theory of gravitation (which however largely coincides with the weak-field slow-motion approximation of General Relativity), it is of no avail to diminish the spatial separation of the two trajectories. It turns out that the ratio between the time intervals provided by a Marzke–Wheeler clock corrected and not corrected for curvature distortions, respectively, is fixed by the intensity of the tidal field present and is independent of the clock's extension.[6] Thus, we cannot approach the undistorted case by building midget clocks, as it were. The difference between the two time intervals decreases under such conditions, to be sure, since both approach zero; but their ratio remains unchanged.

However, there is another option for advancing toward the pure case, namely, operating the clock during continually decreasing periods of time. The ratio of the standard time intervals given by an uncorrected clock and a corrected one critically depends on that period.[7] The shorter the clock is operated the less does the uncorrected clock deviate from its corrected counterpart. This feature allows for the following method of approaching the pure case. We apply the Marzke–Wheeler clock repeatedly during ever shorter periods of time until a further reduction no longer alters the metric time standard obtained. Sooner or later convergence will be reached in any event. If the period of operation vanishes, the time interval supplied by this method coincides with the undistorted true value, irrespective of the intensity of the prevalent tidal forces.

On the other hand, the pure case is certainly inaccessible to direct inspection since it involves a contradiction in terms. After all, it is represented by a situation in which the clock runs zero time and is thus not operated at all. Yet we can approach the pure case by way of a *limiting sequence* of experiments. By continually reducing the clock's period of operation we should be able to decide if we have come close enough to the pure case to get reliable results.

But this is easier said than done, for it seems hardly possible to compare the duration of a second now to the duration of a second some time earlier without recourse to an additional time standard. But this obstacle can be overcome. The general idea is as follows. We know that we have come sufficiently close to the pure case if the time interval as provided by the clock remains unchanged on a further reduction of the clock's period of operation. In order to check this, we measure the duration of one and the same process by two methods; namely, one time using a single integral clock and another time employing several clocks, each being operated only for a short period. That is, after a short period of operation we switch to a new parallel, as provided by the Marzke–Wheeler construction. Then we examine whether the duration of the process under consideration comes out equal on the integral-clock and the split-clock procedure. If it actually does we are dealing with the pure case. If not, we have to repeat the whole maneuver on a smaller temporal scale until, finally, convergent results turn up. The local applicability of the Marzke–Wheeler clock can thus indeed be ensured.

The transportation of the thus established time standard can then be effected as follows. The criterion for the pure state is that the evaluation of a given interval be invariant under clock splitting, as it were. And the upshot of the preceding discussion is that this criterion can be translated into a working experimental set-up. It is then an easy matter to detect the presence of distortions by reversing this approach. We start off from the pure case and gradually enlarge a clock's period of operation until the measured duration of a given process begins to change (as assessed by the just-described method). Next we go over to a freshly constructed parallel and again start an approximation procedure using the limiting sequence method (as just sketched). Afterward the whole approach is restarted. This method solves the difficulty that the Marzke–Wheeler correction procedure was afflicted with (see § 5). The crucial difference to that procedure is that on the present account the metric input requisite to cope with curvature distortions is furnished by the Marzke–Wheeler clock itself. For we are now in a position to ascertain in a non-circular fashion when the switch to a different parallel is in order.

There is, however, one limitation involved in this procedure. It is clear that if the geometry is changing in time we cannot expect to get a convergent series of results in the limit of vanishing period of operation. After all, if the tidal field varies during carrying out the limiting sequence of experiments, their respective outcomes will not approach a fixed value. It is thus to be demanded that space-time structure remains unchanged during the clock's application. Moreover, it is to be borne in mind that the Marzke–Wheeler clock only works satisfactorily if it is operated in free fall motion.[8]

§ 7. It is apparent from this discussion that the Marzke–Wheeler clock constitutes one more instance of a successful resolution of a self-referential circularity by means of external fixing. It will be recalled that it is possible to break up the circle generated by a corrective Reichenbach loop by having recourse to the pure case in which the distorting factor is either absent or prevented from exerting its unwelcome influence. This is made possible by applying additional laws independent of the loop-infected relation. On that basis we are able directly to evaluate the distorting factor and correspondingly to rectify disturbed measurements (cf. II.5.§4). The corrective Reichenbach loop between the evaluation of distant simultaneity and the measurement of one-way transport velocity, for instance, was dissolved in this fashion. It was dissolved, that is, by identifying an experimental set-up (namely, clock transport with vanishing transport velocity) that proves exempt of the disturbing factor in question. This pure or undistorted case is then approached by successive approximation (cf. IV.5.§3).

Quite the same features are present in the Marzke–Wheeler clock. The clock is employed for dissolving the Reichenbach loop between measurements of proper time and curvature evaluations. This is accomplished by turning to a set-up in which the distorting factor ceases to be influential. This does not mean that the distorting curvature does actually vanish under these circumstances (cf. IV.3.§7); it only means that it has stopped exercising its detrimental influence. This peculiarity is analogously found in the Wheatstone-bridge example as discussed in II.5.§1. Moreover, just as in the slow-clock-transport example discussed in II.5.§2, the pure case cannot be realized directly; it rather has to be approached by successive approximation.

What are the principles licensing the conclusion that these successive approximations really approximate the pure case? As just said, we are presented with an instance of external fixing, and so we must ask which laws actually fix curvature. Put otherwise, what justifies the conclusion that curvature no longer becomes manifest in the limit of vanishing period of operation? In fact, this is guaranteed by the properties of Riemannian space-time structure as specified by General Relativity Theory. Apart from somewhat remote effects, the influence of curvature is neutralized in the local inertial frame. It deserves notice that it is the properties of Riemannian space-time structure on which the analysis of the impact of tidal forces on initially parallel paths is based and which likewise help to specify conditions under which this impact vanishes. We are thus again presented with a particularly parsimonious type of external fixing (cf. IV.5.§3). It is the different aspects of the same effect of geodesic deviation that are used for highlighting as well as for avoiding geometric distortions.

I will go into the relevant details only in V.3.§1, but it seems plain already at this juncture that the Ehlers–Pirani–Schild approach in general and the Marzke–Wheeler clock in particular is highly relevant for the problem of assessing the completeness of General Relativity. If a class of phenomena can be singled out as the basis for an "inductive ascent" to a theory, then this theory is certainly able to cope rather exhaustively with these phenomena. Such a class is certainly made up of paradigmatic applications of the theory.

In the case of General Relativity this feature becomes obvious if we compare proper time measurements by means of an atomic clock to such measurements performed by the Marzke–Wheeler clock. The functioning of the former transcends by far the framework of General Relativity; the theory has nothing to say on molecular oscillations. The Marzke–Wheeler clock, by contrast, is based on inertial particle motions and light rays, i.e., on processes that constitute the very core of General Relativity's domain of application. The same holds analogously for the evaluation of spatial intervals. Whereas rigid rods cannot be analyzed within the framework of General Relativity, the radar method is amenable to such an analysis (see § 4). In this vein, Marzke and Wheeler themselves see the upshot of their construction in the conclusion

> that general relativity, in and by itself, provides its own means for defining intervals of space and time [...]".[9]

As mentioned in § 2, the spirit of the Ehlers–Pirani–Schild program flatly contradicts the intuitions characterizing the completeness approach. All the same, the foregoing discussion has made it clear, I suppose, that their technical results can be taken as supporting the completeness of General Relativity Theory. It is seen, namely, that it is possible, relying exclusively on the theory, to identify geodesic trajectories and to construct an instrument, using such trajectories along with light rays, that allows for the empirical determination of metric intervals. The theory thus allows for specifying and analyzing technical tools suitable for establishing its own empirical basis. Furthermore, the theory puts us in a position to cope with geometric distortions, i.e., with the effects of curvature on space-time measurements. We can avoid the detrimental impact of these distortions on our measurements by relying exclusively on the theory (provided that curvature does not change too rapidly). In this way, the conceptual expense necessary for making space-time measurements is kept at a minimum.

So, the (tentative) conclusion is that General Relativity is *complete* with respect to the empirical determination of physical geometry. The *attachment* of indicator processes to the theory's geometric notions can be based entirely on the theory's conceptual apparatus. General Relativity is thus *sufficient* for dealing with geometric measurements. On the other hand, it is not necessary for this purpose since metric intervals may likewise be evaluated by means of atomic clocks. For this reason General Relativity is not strongly complete with respect to space-time measurements. On the whole, then, General Relativity is complete on the theory-level in the sense (3) of the scheme given in II.2.§3: it is sufficient but not necessary for the empirical determination of space-time structure. It is realized, accordingly, that although the Ehlers–Pirani–Schild procedure was conceived for quite a different purpose, it can be reinterpreted so as to provide grist to the completeness mill.

2. COPING WITH DIFFERENTIAL DISTORTIONS, OR: THE ART OF ENTERING THE STATE OF BLISSFUL IGNORANCE

§ 1. The conclusion of the preceding section as to the completeness of General Relativity may appear somewhat premature at the present juncture. After all, it was based on the exclusive consideration of geometric distortions. But it goes without saying that there exist a host of differential effects in addition. So, in fact, the argument doesn't go through so easily (and this is why the conclusion was just qualified as being only tentative in kind).

It will be recalled, in particular, that the Marzke–Wheeler clock (and actually every like-constructed light clock) works with geodesic particle trajectories. We are thus required to identify such trajectories empirically, i.e., to single out freely falling particles. I considered this problem in V.1.§2–3 and ended up with the result that this can be achieved in fact. Yet the consideration was confined to geometrically distorted situations and took no account of differential distortions. So there is an urgent need to look into the matter once more from the differential perspective and to ask whether geodesic trajectories can also be gotten hold of under differentially disturbed circumstances. Put more generally, the question is whether or not it is possible to perform geometric measurements reliably in the presence of differential perturbations.

Expressed this way, it is a question with a past. It was raised by Einstein as an objection to Reichenbach's account of space-time measurements. Reichenbach had claimed that physical geometry can be measured unambiguously after a decision about universal forces has been made (cf. IV.4.§1). Einstein, by contrast, advanced the view that even then the presence of differential distortions may possibly vitiate the empirical determination of metric relations. Let us take a closer look at his argument.

Einstein's point is that differential effects are not that easily detectable and correctable as Reichenbach had assumed. His view is wrapped up in a fictitious dialog between Poincaré (acting as Einstein's spokesman) and Reichenbach. To Reichenbach's claim that physical geometry is uniquely determined once universal forces are conventionally fixed, Einstein offers the following objection on Poincaré's behalf:

> In gaining the real definition [of the rigid body] improved by yourself [by means of differential corrections] you have made use of physical laws, the formulation of which presupposes (in this case) Euclidean geometry. The verification, of which you have spoken, refers, therefore, not merely to geometry but to the entire system of physical laws which constitute its foundation. An examination of geometry by itself is consequently not thinkable.[10]

In order to realize what is at stake here we first have to remove reference to Euclidean geometry. The commitment to the latter merely reflects the predilection of the historical person Poincaré. With this generalization the argument reads: The concept of length already enters into the differential correction laws, namely, into the laws for correcting deformations of the measuring rods that are due to thermal expansion, mechanical stress and the like. In order to determine which rods qualify as rigid and are thus suitable for making reliable length measurements we have to apply correction laws which in turn involve metric quantities. For instance, the law of thermal expansion relates temperature differences to relative length alterations; accordingly, the test of this law presupposes that at some stage reliable length evaluations are at hand. Put more generally, differential corrections can only be performed if the metric is already known. Conversely, the metric can only come to be known by resorting to these correction laws. After all, differential effects of this sort are construed as deformations and not as veridical indications of space-time structure. This reciprocal dependence between the metric and differential distortions induces a circularity that vitiates a separate empirical determination of physical geometry. Whatever the evidence may be, it is always possible to hold fast to one's pet geometry (cf. Grünbaum 1960, 80; Grünbaum 1973, 131–135, 139).

As can be gleaned from the passage quoted, Einstein takes his argument as supporting the thesis that geometry is not testable in isolation but only within the context of additional physical laws. Only the combined system of geometry and physics can be confronted with experience. Einstein is, however, not in the least worried about such test restrictions.

> Why do the individual concepts which occur in a theory require any specific justification anyway, if they are only indispensable within the framework of the logical structure of the theory, and the theory only in its entirety validates itself? (Einstein 1949, 678).

A separate test of geometry is as impossible as it is unnecessary.

It should be emphasized that Einstein's argument does not simply come down to calling attention to Duhem's conventionality thesis. This thesis is concerned with the difficulty of pinpointing the theoretical origin of an empirical failure; and this presupposes of course that the theory arrives at definite, empirically testable predictions in the first place. That is, Duhem's thesis is concerned with Duhemian test problems only (cf. II.2.§3). A case in point is, e.g., how to cope with an eventual conflict between empirical effects of physical geometry and consequences of differential laws. However, Einstein's argument leads in a different direction. As already indicated, it is supposed to support the claim that a *reciprocal dependence* is involved in performing differential corrections of rod-based measurements. The evaluation of a distorting factor (such as changes in temperature or in mechanical stress) requires the identification of the standard of rigidity and vice versa.

This can be reconstructed such that, according to Einstein, a *Reichenbach loop* is present in the differential correction procedure. As a result, it is claimed, this correction procedure is beset with a circularity that prevents us from obtaining any definite geometric data. That is, Einstein's argument expresses the claim that differential corrections actually usher in a *non-Duhemian* test problem (cf. II.2.§3). More specifically, every reliable determination of metric relations demands that differential correction laws be applicable, and this in turn necessitates recourse to reliable metric data. Consequently, differential distortions are in fact *self-referential* distortions of length measurements. And this situation, the argument suggests, is precisely analogous to Reichenbach's universal-force scenario. In the latter, the reciprocal dependence between possible universal influences and the resulting metric is supposed to preclude any separate evaluation of the metric. The Reichenbach loop involved in such a determination makes physical geometry conventional in the first place (cf. IV.4.§3). Accordingly, what Einstein points to here is a differential analog to Reichenbach-conventionality. And this means, geometry remains conventional even after Reichenbach's rule to set universal forces equal to zero has been applied.

§ 2. Unfortunately, Reichenbach's response fails to properly address the problem at issue;[11] it was first dealt with appropriately by Grünbaum. Grünbaum envisages a situation in which only thermal distortions are present. In that case one starts with an arbitrary geometry in the correction law for thermal expansion and determines the geometry with the help of rods corrected by this law. The geometry obtained does not in general coincide with the one employed in the correction. So one uses this measured geometry for effecting the correction a second time and applies the rods corrected in this improved fashion for evaluating the prevailing metric relations. This procedure is to be repeated until an agreement is reached between the geometry that enters into the correction law and the geometry that is obtained through measurements with the such corrected rods. *If* this procedure indeed converges to a result that is independent of the geometry used at the start, it is sensible to identify this resulting geometry with *the* geometry of space-time (cf. Grünbaum 1973, 144–146).

Grünbaum's approach is worth pondering a bit longer. We are presented with a self-referential distortion to the effect that the corrections necessary to extract the prevalent geometry from the differentially distorted data can only be performed if that very geometry is used in the correction procedure. Grünbaum's proposal consists in specifying a *self-consistency approach* to solving this problem. The correct geometry is the one whose application in the correction process yields in turn this same geometry as the result of empirical measurements.

Actually, an approach of this kind is frequently resorted to in physics so as to dissolve apparent circles. A relevant example is the Hartree–Fock method. This method is applied for the calculation of the electric potential generated by the electron shells of an atom. The point is that the potential is determined by the condition that the charge distribution it produces in turn generates this same potential. In order to put this condition into practice, one starts with some potential, calculates on its basis the pertinent Schrödinger wave function of the electronic shell, and then derives from the latter the potential which the corresponding electron distribution gives rise to. If there is no agreement with the potential function used initially, the whole procedure is started afresh, this time inserting the improved potential. The iteration is repeated until input and output in fact coincide.

A similar iterative process also underlies the solution of Einstein's field equations in some instances. This solution involves a reciprocal dependence between the derivation of the metric and the precise shape of the stress-energy tensor figuring in this derivation. The stress-energy tensor expresses the non-gravitational sources of the gravitational field and includes, in particular, the energy and momentum densities of matter and electromagnetic fields, as well as the flow of these quantities, in the region under consideration.[12] In order to obtain the metric from the field equations, the stress-energy tensor has to be calculated. Conversely, this calculation generally involves the metric as realized in the corresponding space-time region. So, a circularity arises that is usually defused by taking advantage of a like-structured self-consistency approach. One begins with some particular metric and calculates using the relevant non-geometric quantities the pertinent stress-energy tensor. Solving the field equations on that basis will yield a metric different than the one employed at the start. Next the entire procedure is redone using the metric obtained as improved input and so on (cf. Ehlers 1973b, 113). Relying on this iterative method we finally end up with a self-consistent result.

§ 3. Let us take a closer look at the logical relations that hold between the two quantities involved in a successful self-consistency approach. It is apparent from the outset that they cannot be connected to one another by an ordinary Reichenbach loop. If only one hypothesis were relevant here, then the evaluation of the quantity q_1 via the quantity q_2 could not possibly diverge from the initial q_1-value. In usual Reichenbach loops self-consistency is achieved automatically, and this is precisely what makes them epistemically suspicious in the first place. In the scenarios under consideration here, by contrast, the nomological network appealed to entails exactly one evaluation for q_1 and q_2. Here, only one solution is allowed whereas in Reichenbach loops no value is excluded.

What is the logical feature that is responsible for this noteworthy difference? The distinctive feature is that a dependable self-consistency approach relies on at least two independent hypotheses. To wit, hypothesis H_1 is employed for inferring a value of q_2 from a value of q_1, and a different hypothesis H_2 is used for calculating q_1 from q_2. Let me flesh out this abstract scheme by reconsidering the two examples presented above. In the Hartree–Fock example the two relevant quantities are the electromagnetic potential and the charge distribution of the electrons. The inference from the former to the latter proceeds essentially by solving Schrödinger's equation. The reverse inference, by contrast, is governed by Maxwell's equations. Analogously, the transition from the stress-energy tensor to the metric relies on Einstein's field equations while the backward calculation is quite independent of these equations and is based instead on generally covariant expressions of energy and momentum density.

This scenario thus bears a superficial resemblance to external fixing but the difference is that in contrast to the latter the two quantities are fixed reciprocally (rather than one of them being fixed independently and the other computed via the loop hypothesis) (cf. II.4.§4). In contradistinction to external fixing, we are presented with two relations connecting the two relevant quantities. On that basis we can try a procedure of successive approximation by reciprocal adaptation. And if the pertinent relations are structured such that they really single out one definite evaluation, then the results obtained will converge toward this distinguished value. In that case, the reciprocal dependence is converted into a helical procedure, as it were, that finally provides us with the unique evaluation of the two quantities involved as implied by the corresponding system of hypotheses. A nomic net that allows for such a reciprocal adaptation I call a *Grünbaum helix*. A Grünbaum helix is thus characterized by two features. First, there are (at least) two hypotheses relating two quantities to one another; second, the relation established between these quantities is of such a kind that a unique solution exists.

This consideration shows that not all self-referential distortions arise from Reichenbach loops. As explained in II.5.§3, corrective Reichenbach loops spawn self-referential distortions; but as we realize now, the latter may also originate from other sources. Self-referential distortions may be generated as well by a Grünbaum helix. Moreover, it is easily seen that a Grünbaum helix constitutes a possible complication of potential Einstein–Feigl completeness on the theory level (namely, if the two hypotheses in question are part of the same theory). In all cases hitherto discussed, the emergence of non-Duhemian test problems critically depended on the fact that only one single hypothesis was taken into account. That is, it was completeness on the hypothesis level alone that seemed to engender non-Duhemian test problems (along with the concomitant threat to testability). We now see that even completeness on the theory level is not always epistemically innocuous, although we realize at the same time that if the relevant complications really exhibit the structure of a Grünbaum helix, testability can be reestablished by means of a self-consistency approach.

§ 4. Now the crucial question is whether Einstein's differential correction circularity really constitutes a Grünbaum helix. The answer is affirmative indeed. There is, first, a reciprocal fixing of the metric and a differential distortion and this relation can be transformed into an adaptive helix. The inference from the distorting factor (such as temperature variations) to the undistorted metric is based on the respective differential correction law while the evaluation of the distorting factor relies on the observation theories of the instrument employed along with geometric theorems (that allow for the calculation of lengths, areas and volumes from a given metric tensor). So the first condition is actually met.

In principle, the same applies to the second one: A convergent series of results will often show up. However, there are some qualifications to be heeded. It is immediately clear that convergence will only emerge if there exists an unchanged value these results can converge toward. This means, first, that in order for the method to work successfully, the curvature of the region under scrutiny must be neither time- nor position-dependent.[13] As a matter of fact, however, this is not always the case. General Relativity associates spatiotemporal curvature variations with a number of gravitational effects. Second, this method only supplies us with sensible outcomes if the distorting temperature variations remain unchanged during its application. It is clear that if the strength of the perturbing influence changes in the course of the correction process, we cannot expect to obtain a convergent series of results. Hence, this approach relies on the condition that the spatial pattern of distortions does not change in time. Only on that condition this pattern can be extracted from successive measurements.

The situation is further complicated by the obvious need properly to procure the necessary empirical input. This input, however, is sometimes hard to come by. Consider the case of inelastically deformed rods. In order to carry out the reciprocal adaptation procedure reliably it is requisite that the spatially variable intensity of the deforming influences be faithfully reflected in corresponding changes in the length of the rods. This requirement is violated if there occurs a "hysteresis" due to inelastic deformations. In that case the rod does not regain its original length after the distorting influence has ceased but rather retains a somewhat altered length. Consequently, the rod would indicate the presence of a distortion where in fact none exists.

The point is that judging whether the condition of perfect elasticity is met demands recourse to reliable length measurements which in turn were supposed to be established by the whole procedure. And this shows that there is another differential correction circularity involved here whose dissolution would call either for the identification of ideally elastic rods through external fixing (i.e., independent of any length measurements) or for a second application of the self-consistency procedure. I am not aware of any procedure suitable for implementing either option.

Anyway, the gist of Grünbaum's analysis consists in the demonstration that the apparent differential correction circularity in reality exhibits the structure of a Grünbaum helix and thus lends itself to a treatment via reciprocal adaptation. And the present consideration leads to the conclusion that this treatment is in principle fit to defuse the situation under discussion. On the other hand, the method is only applicable if the perturbing factors remain unchanged, and it moreover involves technical complications that make it difficult actually to get hold of the unique value implicitly contained in the nomological network at hand. The method is thus unsuited as a general means for disentangling differential distortions and geometric effects. Grünbaum's ingenious application of the self-consistency approach to rigid rods, useful as it is in some cases, is not generally applicable.

§ 5. Since we came up dry in actually accommodating differential distortions in a general fashion, we might get lucky if we simply discount them. Why bother about distorted cases? Why not simply forget about them? After all, reportedly there exist instances of blissful ignorance. Implementation of this nonchalant strategy requires the identification of differentially undistorted rods without availing ourselves of differential correction laws. On several occasions we have already applied this strategy of realizing the pure states. So let's attempt another application of this tried and tested remedy.

Fortunately enough, there exists an option for freeing ourselves from the irksome particulars of the correction laws at issue. Namely, whatever these laws may be like they all agree on one thing: If there is no differential distortion there is no need for a differential correction. If we manage to secure the absence of differential perturbations without invoking rods already known to be rigid we are done – or so it seems.

Translation of this strategy into a working procedure has again been tried by Grünbaum. He proceeds along the following lines. What we can do even on our fairly restricted conceptual basis is judging whether or not the length of a given rod is equal to that of a second rod placed next to it in the same direction. Local congruence can be established non-metrically. And this capacity suffices for ascertaining the absence of perturbations. If every pair of rods of different chemical composition and in local congruence always remain in local congruence when being jointly transported to different locations, then this reliably indicates that the region under consideration is perturbation-free. In that event it is safe to conclude that the geometric relations obtained truthfully indicate the prevailing space-time structure (cf. Grünbaum 1973, 136).

But this procedure suffers from a serious shortcoming. As already mentioned in IV.4.§2, universal causes may have differential effects. An awkward cause in point is tidal forces. These substance-unspecific forces tend to deform the rod and to alter its length. But since the actual deformation of the rod is the joint effect of the tidal force and the substance-specific force of cohesion, the resulting deformation depends on the material employed. As a result of the different elasticity of different materials, changing tidal forces may induce responses of different strength in different materials. This implies that chemically distinct rods may not remain locally congruent under transportation because curvature variations occur and in spite of the fact that the region under consideration is free of differential perturbations.

Consequently, Grünbaum's perturbation test may be falsely positive. The test is not only sensitive to differential distortions but also to changing curvature (or changing tidal forces). It may indicate the presence of differential distortions where there are in fact none. Grünbaum's test is indeed sufficient to ascertain that a certain region is perturbation-free (namely, if local congruence is retained). The reverse, however, does not hold. One cannot reliably detect the presence of perturbations, and one cannot make sure, consequently, that all perturbation-free regions are correctly identified.

There is a second limitation involved in Grünbaum's approach. Suppose that the perturbation test is rightly positive and let's ask what can be derived from that result. The answer is: precisely nothing. More specifically, we cannot extract the prevailing geometry from distorted cases. That is, we cannot derive the geometry present in distorted situations from the geometry realized in the undistorted ones. Such a derivation would require a comparison between the distorted and an undistorted region, and in that case we can never rule out that the geometry is actually different in both regions. So the transition from a perturbation-free state of affairs to a distorted one is blocked here. It is clear, however, that differential distortions are almost ubiquitous in our universe. Accordingly, the strategy of realizing pure cases, when applied to the rod scenario, is of only limited help.

§ 6. It must thus be admitted that Einstein's circularity argument indeed possesses some force and cannot be dismissed easily. However, the argument as well as the hitherto discussed options for defeating it all referred to measurements by means of rigid rods. But if we regard the situation from the angle of completing space-time theory, rod-based measurements are less than ideal anyway and are better supplanted by methods relying on particle trajectories and light rays (cf. V.1.§7). So let's try this route and consider measurements performed along the lines discussed in V.1.§2, §4. What makes this route appear particularly promising is the fact that neither temperature variations nor mechanical stress nor intricate substance-specific responses to substance-unspecific forces exert their unwholesome influence there.

Actually, however, this hope is also defeated since the trouble is merely shifted to a different level. A differential distortion that may affect particle motion is electric charge. If the particles used for probing space-time structure are charged, the influence of possibly present electromagnetic fields on their trajectories has to be evaluated and (as the case may be) corrected. Such corrections can be carried out by applying Lorentz's force law. The problem with that procedure is that in the generally covariant formulation of that law the metric appears: $f^i = qg^{ik}F_{kl} \times (\mathrm{d}x^l/\mathrm{d}\tau)$ (with f^i denoting the correcting Lorentz force and F_{kl} the electromagnetic field strength tensor) (cf. Weinberg 1972, 125; Wald 1984, 69). As can be gleaned from this equation, there is a reciprocal dependence between the metric g^{ik} to be determined and the correcting electromagnetic force f^i. Calculating this force from Lorentz's law requires knowledge of the metric, and conversely, the metric can in general only be determined by applying this force law to differential trajectory deformations and thereby rectifying them. This situation thus apparently leads into a circle that parallels the one in Einstein's rod-scenario: In order to reliably measure the metric we have to carry out corrections, that in turn make use of the metric. This means that, from the perspective of the trajectory method, distortions induced by lack of neutrality are in fact *self-referential*. Correspondingly, Einstein's differential circularity problem has a general bearing on space-time measurements.

Now the question emerges naturally whether this problem-situation can be transformed into a Grünbaum helix? The discouraging answer is that it can't. Such a rescue maneuver would call for a second hypothesis. But the only *prima facie* candidate, namely, Maxwell's equations turn out to be a failure for the simple reason that they do not establish a connection between the two relevant quantities (i.e., the metric and the Lorentz force). The displeasing conclusion is that we are confined to only one hypothesis and that we are consequently faced with a Reichenbach loop. This being the case, we could try to launch a relief operation by externally fixing one of the two quantities involved. One might again think of Maxwell's equation as a means for establishing the metric independently of Lorentz's force law. The properties of light propagation, as encoded by Maxwell's equations, might appear to provide the metric without appeal to particle motions. But in fact they don't; light propagation is insufficient for singling out the metric (cf. V.1.§2). The alternative option is to try to obtain the complementary loop-constituting quantity independent of the loop-infected relation. This requires external fixing of the Lorentz force, i.e., evaluating it without recourse to

Lorentz's force law. A natural way to this end is a direct measurement of the acceleration values of charged particles. But such measurements obviously presuppose a spatiotemporal metric. So, neither Maxwell's equations nor the evaluation of charged particle accelerations provide a suitable basis for external fixing. Accordingly, the corrective Reichenbach loop involved in the application of Lorentz's force law is here to stay.

§ 7. Thus only one possibility remains open to us. This last resort scheme consists in realizing pure instances in the trajectory scenario. Our task is, then, to identify differentially undisturbed particle trajectories without relying on metric procedures. And this task obviously comes down to singling out geodesically moved particles. If this can indeed be accomplished, deviations from these geodesic paths are to be attributed to the influence of differential perturbations.

Now we are in the fortunate position of being already familiar with a procedure appropriate for this purpose. As explained earlier, the Ehlers–Pirani–Schild criterion for geodesic motion supplies us with an effective test for the identification of geodesics based on their internal properties (cf. V.1.§2–3). And in fact this is all we need here. The ability to identify geodesics amounts, namely, to the capacity of securing empirically the absence of differential distortions. This implies that we are now in a position to reliably set apart the motions of neutral and charged particles. Only the former, and not the latter, pass the geodesics test. It is then an easy matter to evaluate electromagnetic perturbations. For that purpose we only have to compare the paths of neutral particles to those of charged ones in the same space-time region. The latter's deviations indicate the intensity of the distortions present which can consequently be quantified and corrected. Unlike the rod scenario, the comparison between the undistorted and the distorted case can here be effectively carried through.

It is worth noting that this procedure is not restricted to electromagnetic distortions. It is not confined to situations in which the nature of all possibly occurring differential distortions is known beforehand. The reason is that the pure state is not identified by a stepwise exclusion of distortions but rather by means of a direct operational criterion. After an empirically applicable criterion for geodesics is at hand, we can simply count every particle whose motion deviates – whatever the cause may be – from the thus distinguished paths as distorted. In order to carry out this procedure we need not know in which way its motion is distorted. This is why the selection of undistorted instances represents the highly sophisticated art of remaining blissfully ignorant of unwelcome things.

This shows that physical geometry can be obtained, even in distorted cases, without making use of additional physical correction laws. That is, pure instances can be selected or self-referential distortions can be avoided without loss of generality. This works because we can arrange things such that the perturbing influences have no effects. Neutral (small, non-rotating etc.) gravitational monopoles follow unflustered their geodesic paths, regardless of all turmoil on the differential level. This peculiarity has no analog in the rod-based method which is why it is of no general applicability.

On the other hand, the method of selecting undistorted trajectories does not represent a truly comprehensive solution to Einstein's differential circularity problem for it evidently presupposes that differentially undisturbed particle paths are empirically accessible to us. If, for instance, only charged particles existed, no general way out of Einstein's problem could be specified. Since, however, nature is kind enough to supply us with the relevant pure states, the problem can successfully be tackled. Accordingly, the conclusion stated at the end of V.1.§7 as to the completeness of General Relativity with respect to the empirical determination of physical geometry turns out to be justified in the end.

3. COMPLETING PHYSICAL GEOMETRY, OR: THE ART OF ENTERING THE STATE OF SPLENDID ISOLATION

§ 1. In an axiomatization of a theory a certain subset of all the propositions making up this theory is placed at the top and the remainder is derived from this distinguished subset. The distinguished propositions are framed using a likewise distinguished vocabulary. They are framed in terms of "primitive descriptive concepts" (drawing on a logico-mathematical framework where necessary). From these basic concepts the remaining conceptual apparatus of the theory at hand is built up with the help of explicit definitions so that in the end all the claims constituting the full-grown theory can be derived from this restricted conceptual-propositional basis.[14]

It is of high philosophical significance what kind of concepts are taken as primitive and what kind of propositions are accepted as basic. Reichenbach introduces, and at the same time recommends, a type of axiomatization in which the primitive concepts refer to observable phenomena and in which, consequently, the propositions placed at the top express immediately observable facts. This type Reichenbach calls *constructive axiomatization*; and his reason for favoring it over approaches which proceed by singling out more abstract and theoretically far-reaching principles as fundamental is that in constructive axiomatizations the axioms are more directly amenable to empirical scrutiny. While Reichenbach acknowledges that even constructive axioms can only be verified with the help of observation theories (in the sense specified in I.2.§5), he still insists that they are more elementary than higher level axioms such as variational principles. They are more elementary in the twofold sense of being qualitative in kind and of requiring less sophisticated technical means for their establishment. Bearing in mind the qualifications, constructive axioms can thus be interpreted as capturing elementary empirical facts (cf. Reichenbach 1924, 2–5).

The Ehlers–Pirani–Schild program can be construed as being aimed at a sort of constructive axiomatization of the geometric structure of space-time (as specified by General Relativity Theory). Ehlers–Pirani–Schild wish to clarify, first, which facts justify the assumption that the space-time structure is pseudo-Riemannian in kind and, second, by which means the space-time geometry prevailing in particular regions can be empirically ascertained.

> We wish to show how the full space-time geometry can be synthesized from a few assumptions about light propagation and free fall. [...] [We] attempt to derive the conformal, projective, affine, and metric structures of space-time from some qualitative [...] properties of the phenomena of light propagation and free fall that are strongly suggested by experience. Not only the measurement of length but also that of time then appears as a derived operation (Ehlers/Pirani/Schild 1972, 65).

Commitment to simple, qualitative experiences as basic precludes application, first, of theorems of General Relativity itself and, second, of theories foreign to the framework of physical geometry in order to derive measurement procedures for spatiotemporal quantities.

To systematize and generalize somewhat, a constructive axiomatization deliberately forgoes invocation of two types of conceptual resources. It first renounces drawing on the theory to be axiomatized and it second restricts as much as possible the use of observation theories. The second condition can be stated as a regulative principle not to apply any observation theory in the construction, i.e., as a principle we want to live up to although, strictly speaking, we cannot. If we are urged not to resort to any observation theory in the construction, it follows that we are not allowed, in particular, to employ any observation theory different from the theory to be constructed.

It is this latter requirement where constructive axiomatizations and the completeness approach meet. The requirement of non-interference addressed to additional, external observation theories provides the common ground between the two approaches. For instance, basing spatiotemporal measurements on rigid rods and atomic clocks appears equally unsatisfactory from both angles. What makes a particular rod suitable as standard of rigidity is determined by its molecular constitution. So, rod-based measurements can (if at all) only be analyzed within quantum chemistry; they clearly transcend the range of spatiotemporal concepts. The same goes analogously for atomic clocks. Molecular oscillations have no place in physical geometry, and General Relativity has nothing to offer as to why these oscillations exhibit their unrivaled frequency stability.[15] Using quantum theory as an observation theory for physical geometry is rejected by both approaches under discussion.

Because of this shared stance the results of constructive axiomatizations are of some use for the implementation of the completeness approach. Phenomena suitable for an Ehlers–Pirani–Schild-type construction from scratch do not require for their interpretation a great deal of external theoretical input. They are thus likely to be treatable, at least to a great extent, by the relevant theory itself. It is the shared preference for elements not infected by extraneous agents which makes Ehlers–Pirani–Schild-type phenomena (or for that matter, phenomena covered by constructive axioms) promising candidates for supporting the completeness of the pertinent theory.

§ 2. This agreement does not, however, extend to the first constructive non-interference postulate, i.e., to the demand that the theory to be axiomatized stay out of the axiomatization process. Here the constructive and the completeness approach part company. For the latter it is desirable that the theory at issue looms large in the analysis of the respective measurement procedures. If we ask in the constructive framework why a (properly operated) light clock is a suitable means for evaluating space-time intervals we are referred to a selection of simple experiences. It is such elementary facts and relations that the measuring instrument is supposed to be based on, and it is in this sense that space-time measurements are considered "derived operations" (as the quote in § 1 indicates). The completeness advocate, by contrast, would respond to the question by spelling out the observation theories involved in the measurement procedures and, in particular, by looking for nomic ingredients of the explanatory theory (i.e., General Relativity) itself. Measurement procedures are derived operations in that they are consequences of more comprehensive theories.

The first thing to be clarified is the role the theory to be axiomatized plays in the constructive axiomatization. It is clear that this theory cannot entirely be kept out of the plot. Rather, the theory enters it in a veiled fashion, namely, by supplying *criteria of adequacy* for the measuring procedures. And the theory does so by defining the pure states the instrument is intended to register. It provides the criteria for selecting facts as basic. After all, what is the justification for regarding the motion of test particles in a gravitational field as the realization of the pure state while considering the motions of extended bodies or charged bodies as distorted. What is operative here behind the scene is the knowledge that by singling out the former motions, but not the latter, as perturbation-free a consistent and comprehensive account of all relevant motions can be given. The theory thus serves to cast the characters, as it were: It stipulates what is to qualify as an ideal case and what as a distorted one. Only because we already know General Relativity, can we feign to be ignorant of it and construct it from scratch.

It must be admitted, on the other hand, that General Relativity does not figure as an observation theory in the Ehlers–Pirani–Schild axiomatization. Their constructive axioms can indeed be verified empirically without getting involved in the higher-level apparatus of General Relativity. But the ability to keep General Relativity out in no way compels us to do so. So let me now set out to reinterpret their results so as to integrate their approach into the completeness project. For that purpose, the machinery of General Relativity has to be brought into play full-scale. I begin by asking what the field equations actually entail as to the motions of freely falling particles and the paths of light rays.

§ 3. In Newtonian mechanics, the field equations for the gravitational field (i.e., Newton's law of gravitation or its local formulation, viz. Poisson's equation) are independent of the pertinent equation of motion. The latter cannot be derived from the former but has to be postulated separately. In General Relativity, however, matters are different. Some important features of particle trajectories and paths of light rays actually follow from Einstein's field equations of the gravitational field. Outstanding among them is, first, the geodesic nature of free fall motion and, second, the fact that light rays represent null geodesics. This is to be understood roughly as follows.

Local conservation of energy-momentum is a direct consequence of Einstein's field equations, and from this principle it can be derived that a test particle pursues a geodesic if it moves uninfluenced by any external, i.e., non-gravitational, force. Massive bodies tend to change the prevailing geometry so that such a body's path will be non-geodesic with respect to this "background geometry"; that is, with respect to the geometry set up by the surrounding bodies. Similarly, extended objects will likewise depart from a geodesic path because geodesic deviation (induced by the possibly present space-time curvature) alters the trajectory of an object of that kind. Furthermore, if a particle is rotating or possesses higher gravitational multipole moments, it, too, is deflected from a geodesic path. Effects of this sort represent geometric distortions of particle motions (cf. V.1.§1). Making allowance for these qualifications, test particles still do move on geodesics with respect to the background geometry, and this remarkable feature follows from the laws governing the behavior of the gravitational field (cf. Misner/Thorne/Wheeler 1973, 476–479; Stephani 1980, 87–89).

In addition, the propagation of light in curved space-time is described by the generally covariant form of Maxwell's equations. Light rays can be characterized by the vectors lying perpendicular to the respective light wave's planes of constant phase. It can be shown, then, relying on Maxwell's equations, that these wave vectors are null vectors and that they obey a geodesic equation of motion. In sum, light rays are null geodesics (cf. Misner/Thorne/ Wheeler 1973, 572–576; Landau/Lifshitz 1976, 293–294). Furthermore, Maxwell's equations can in part be derived from Einstein's field equations. More precisely, given the latter equations and given the stress-energy tensor of the electromagnetic field, the so-called second group of Maxwell's equations follows (cf. Misner/Thorne/Wheeler 1973, 471–473; Landau/Lifshitz 1976, 331).

The conclusion is, accordingly, that the geodesic nature of free fall motion and of light propagation is in important respects a consequence of the fundamental principles of General Relativity. It will be recognized that this achievement constitutes an essential step toward the completeness of that theory. The theory succeeds in singling out phenomena corresponding to its basic theoretical notions. It manages to pass on from its central equations to empirically tangible phenomena without having to draw extensively on additional auxiliary theories. On the other hand, we are not yet done. Completeness calls for the derivability of a feasible procedure that allows for tracking down the processes or entities shown to be theoretically significant. The mere self-sufficient derivation of empirically testable results scores a point on the parsimony scale but falls short of actually establishing completeness.

This is where the Ehlers–Pirani–Schild scheme enters the stage. General Relativity entails that test particles move along geodesics. So, in order to translate this theorem into a working measuring procedure we have to be able to identify either geodesics or test particles unambiguously. The Ehlers–Pirani–Schild criterion for geodesic motion indeed provides a means for fulfilling the first demand: It reliably supplies us with the requisite geodesics even under unfavorable conditions (cf. V.1.§2–3; V.2.§5). Moreover, once we have succeeded in extracting geodesic trajectories we can apply the Marzke–Wheeler clock and measure out the prevailing metric relations (cf. V.1.§4–6).

Viewed from the perspective of the completeness project, the question to be asked now is: What are the theoretical principles that justify invoking the geodesic criterion and the Marzke–Wheeler construction? We have to spell out which laws serve to explain why these two procedures are adequate. In other words, we have to unearth the relevant observation theories. First and foremost among them is the principle of equivalence (cf. IV.3.§1). It is by virtue of this principle that we consider the motion of test particles in a gravitational field as perturbation-free or geodesic. It is this principle that licenses the geometrization of gravity in the first place. A second ingredient of the procedures mentioned is the local applicability of Special Relativity. It is required that in the local inertial frame Special Relativity holds, i.e., that the metric is Minkowskian. In the Marzke–Wheeler clock this condition is appealed to twice; first, in the construction of a parallel to a given straight line, and second in the use of reflected light rays for fixing equal time intervals. The latter stipulation invokes the constancy of the speed of light. Following out the consequences of these two conditions leads one naturally to the assumption that the actual space-time structure is pseudo-Riemannian in kind.

The pseudo-Riemannian space-time structure thus explains why the geodesic criterion is adequate and why the Marzke–Wheeler clock works appropriately. In other words, it can be derived from the principle of equivalence and the local validity of Special Relativity that the geodesic criterion really serves its intended purpose and that the Marzke–Wheeler clock really furnishes a reliable measure of metric relations. These laws allow for an analysis of the trustworthiness of the two devices and thus justify their application. Since both assumptions form an essential part of General Relativity we may conclude that the theory is *Einstein–Feigl complete* with respect to determining the geodesic as well as the metric structure of space-time. The evidential basis of General Relativity can be ascertained using procedures that are built-in to it. This is to be expressed such that General Relativity is *sufficient* for attaching indicator procedures to its fundamental geometric notions (cf. V.1.§7).

§ 4. This conclusion holds at least under ideal conditions. For up to now we have only considered the attachment of indicator phenomena to the pure and undistorted theoretical states. So, in order to secure the conclusion we have to turn to distorted cases and examine General Relativity's capacity to perform the necessary corrections in a self-sufficient fashion. Let's begin by focusing on the influence of geometric distortions on particle trajectories. As already indicated in § 3, only test particles follow a geodesic path; it is immediately clear, however, that test particles are somewhat idealized objects and that they are not abundant in our surroundings. Remember, test particles are point masses with negligible mass values. So, we are in need of theorems that apply to more tangible entities such as planets, for instance. Planets are heavy, extended and rotating bodies, and they possibly possess an asymmetric mass distribution. Non-ideal cases of this kind obviously exist all over the place, and the question is whether General Relativity can deal with them appropriately. Does the theory entail anything about the paths of planets?

It does in fact. It is actually possible to accommodate geometrically distorted motions within General Relativity. In particular, the theory allows for the derivation of an equation of motion that does indeed hold for bodies endowed with a non-vanishing mass, a non-vanishing intrinsic angular momentum and gravitational multipole moments (cf. Ehlers 1973b, 102–105; Stephani 1980, 89–90). One obtains, as anticipated in § 3, that the trajectories of such bodies are non-geodesic. The interaction of the mentioned body properties with the tidal forces due to space-time curvature drives such bodies off their geodesic courses – unless the magnitude of the distorting influence is negligible (as it actually turns out to be in the case of the planets).

The point is, then, that the magnitude of the distorting effects on a body's motion can be calculated and thus be taken into account. This holds at least if the metric is given so that the equation of motion can be applied directly. If it is not given, we have to resort to the self-consistency procedure sketched in V.2.§2. In this way it is in principle possible to evaluate and correct the effects of geometric distortions on the motions of non-ideal bodies. The metric can thus be extracted from distorted data via the field equations (using body parameters in addition). We are thus not compelled to resort to the Marzke–Wheeler clock which requires that geodesic, i.e., undistorted, motions be realized.

Still, even to the latter procedure General Relativity does a useful service by highlighting the disturbing influence of curvature on the clock's running. The theory shows that and how curvature vitiates the parallelism of the two particle world-lines that make up the clock; and it gives a quantitative evaluation of this influence – provided that the metric is given. That is, if the metric is at hand, General Relativity predicts the clock readings as affected by non-vanishing curvature, and it says by what measure they are wrong. We can calculate how curvature expresses itself in erroneous clock readings. By contrast, in order to obtain the metric using the Marzke–Wheeler clock, recourse to the undistorted case is inevitable (cf. V.1.§5–6). Accordingly, although the curvature effects perturbing the operation of a Marzke–Wheeler clock can be analyzed in a complete fashion, the clock itself furnishes the evidential basis for the corresponding correction procedure only if undistorted trajectories are available.

The same holds analogously on the geodesic level. It follows from the general applicability of the geodesic criterion (cf. V.1.§2–3) that we are indeed able to single out undistorted particle motions. The discussion of differential perturbations has made it clear that if we observe a trajectory that deviates from the corresponding geodesic we can attribute the difference to the presence of distorting factors and thus rectify their influences (cf. V.2.§5). But this constitutes a correction procedure in a Pickwickian sense only. It supplies us with the pure state on the condition that the pure state is already known.

This shows that the capacity to perform corrections is severely limited if we confine ourselves to methods that can be specified on the basis of constructive axioms. On the latter basis we can reliably separate the wheat from the chaff, as it were, but we are unable to recycle the chaff; we rather have to discard it altogether. Put less figuratively, we can set apart geodesic and non-geodesic motion but there is no (non-Pickwickian) way from the latter to the former. In order to achieve more than the selection of pure instances we have to specify the nature of the distortions present and bring to bear the relevant theories. As regards distortions due to charge, for instance, we have to make use of full-scale electrodynamics for carrying out the necessary corrections. Since, however, electrodynamics is in part derivable from Einstein's field equations (see § 3), these corrections can in part be performed in a complete fashion. The conclusion is that General Relativity is in large measure sufficient for performing the corrections requisite in the empirical determination of spatiotemporal quantities.

On the other hand, there exist distortions which are in no way rectifiable on the basis of this theory. Consider curvature distortions induced in an atomic clock. If the tidal forces associated with curvature exceed a certain critical level they deform the macroscopic structure of the clock to such an extent that the clock ceases to run appropriately. At a still higher level the oscillating molecules themselves are deformed with the result that the frequency stability is lost (cf. Misner/Thorne/Wheeler 1973, 396). Evidently, effects of this sort are not covered by General Relativity Theory. While the theory is necessary for evaluating the intensity of the distorting tidal field, it contributes nothing to the clarification of the detrimental influence of this field on the clock's appropriate operation. Consequently, the correction procedure proper proceeds independently of General Relativistic theorems and has to be based instead on technical mechanics and quantum mechanics. But this is a far cry from a complete correction. In the rectification procedure for curvature distortions in atomic clocks General Relativity is virtually absent.

The same holds analogously for the use of ordinary rigid rods. It is true that the intensity of deforming tidal forces is specified by General Relativity. Yet the correction procedure has to take the response of the respective materials into account, and this response is essentially governed by molecular cohesive forces (cf. IV.4.§2, V.2.§5). Furthermore, if we wish to correct differential distortions by bringing to bear the Grünbaum helix expedient (cf. V.2.§3), we have to resort to differential correction laws (such as the law of thermal expansion) that cannot be considered part of General Relativity. Again, the theory is conspicuous by its absence.

From this we may draw the following conclusion as to the Einstein–Feigl completeness of General Relativity with respect to corrections necessary for making reliable measurements of spatiotemporal relations under distorted circumstances. There is one track from the perturbed evidence to the pure theoretical state on which the theory is able to proceed in splendid isolation. It is nearly sufficient and certainly necessary for performing the relevant corrections. This applies to all procedures that are based on particle (or body) trajectories and paths of light rays. There is another such track, however, exemplified by the application of rigid rods and atomic clocks, on which the theory is of virtually no help. Accordingly, with respect to corrections the theory is complete in the sense (3) of the general scheme given in II.1.§3: It is (nearly) sufficient but not necessary for rectifying perturbed spatiotemporal evidence. This result reproduces the assessment of the theory's completeness with respect to indicator attachment (cf. V.1.§7, V.3.§3). General Relativity is sufficient but not necessary for establishing and correcting procedures suitable for evaluating empirically spatiotemporal quantities.

4. HOW COMPLETENESS IS POSSIBLE, OR: THE ART OF IMPLEMENTING A TOUGH CONCEPTUAL AUSTERITY PROGRAM

§ 1. The somewhat technical explorations in the preceding sections have led to the conclusion that General Relativity is complete to a remarkable extent. The task left is to gain a little more clarity about what precisely is the property that we can attribute to General Relativity in this noteworthy degree. So in this section I will expand somewhat on the *semantic* and *methodological* aspects of completeness. For this purpose it is advisable to step back a little from the technical machinery of General Relativity. The intricacies of the gears and levers may prevent us from getting an overall picture. Essentials sometimes come out more clearly from a distance. "Appalachian Spring" was composed in California. Thus, I will try to elucidate a bit the general intuitions that characterize the completeness program. In order to heighten the profile of the program I contrast it with the semantic and methodological intuitions underlying constructive axiomatizations.

Constructive axiomatizations attempt to ground all the core assumptions of a theory on the bedrock of simple experiences. That is, a construction is envisioned that proceeds on this firm foundation without gaps or circles. The construction is *linear* in the sense that no part of the developed theory is employed in building it up. This methodological commitment implies that the meaning of the terms figuring in the theory is likewise to be based on the simple experiences resorted to in the construction. In particular, it is solely experience that supplies referents for the concepts employed. This consideration shows that constructive axiomatizations are committed to an operationalist approach which is in turn tied up with a verificationist semantics (cf. I.1.§1). On this understanding, the theory makes no contribution to the clarification of the concepts it consists of. Meaning derives from the relevant evidence alone. Hence, in addition to methodological linearity there is a commitment to semantic linearity. The definition of the requisite concepts must come first and it must proceed from basic experiences alone; subsequently, factual statements are made with the so defined concepts.

Completeness is not in the same ballpark, neither semantically nor methodologically. In order to see this more distinctly we ask what completeness amounts to on the semantic level. The answer is, roughly speaking, that a complete theory is able to connect by its own lights empirically directly accessible referents to its own concepts. So, our task is to identify a semantic account that proves helpful for elucidating this capacity.

The double-language model (as sketched in I.1.§2), certainly does not. The reason is that on this model scientific theories are represented by uninterpreted formalisms in the first place which receive their interpretation through the correspondence rules adduced to the formalism (cf. I.1.§3). In this semantic framework the demand that a theory associate referents to its own concepts means that a formal algorithm is to entail the relevant correspondence rules. And this is tantamount to the demand that an uninterpreted structure be able to generate its own semantic interpretation.

However, the self-sufficient generation of a semantic interpretation from a formal algorithm comes down to the analog of a veritable abiogenesis. And this is no less excluded in semantics than in biology. Formal structures are always multiply interpretable (cf. II.2.§1); and this means that we can always attach more than one set of referents to the abstract symbols occurring in a formalism. Consider the example of the abstract equation describing damped oscillations in general. In this equation the temporal evolution of a formal variable is related to certain likewise formal parameters. The equation thus represents an uninterpreted formalism. It can be applied to a host of different situations by tying its abstract symbols to appropriate physical quantities. It may, for instance, account for the oscillations of a coil spring in terms of the values of the oscillating mass and the elastic constant of the spring etc.; or it may explain the temporal development of current intensity in terms of, e.g., the self-inductance and the capacity of the relevant electric circuit.

None of these different interpretations can be distinguished on the basis of the formal equation alone (cf. Carrier 1991a, 48, 50). We are faced with an embarrassment of riches that keeps us in a state of semantic ambiguity and thus prevents us from reaching any definite interpretation. So, the goal of extracting substantive content from an abstract formalism unaided by anything external to this formalism is ill-conceived and incoherent. There is no way to squeeze semantic juice out of a non-semantic orange. If understood this way, completeness is impossible to accomplish.

§ 2. If we wish to fare better we must proceed differently. An obvious option is clarifying the meaning of a concept using other, already interpreted concepts. And what appears particularly promising is determining the meaning of a concept via its use in interpreted laws and theories. This method represents the very pivot of the *theoretical context account* of meaning. The context account operates with interpreted laws and theories so that there is no procreation of meaning out of form involved here. According to this approach, the meaning of a scientific concept is fixed by the network of laws it belongs to. Meaning derives from theoretical integration (cf. I.2.§3).

We cannot stop here, of course, for the natural follow-up question is in which way the meaning conferring laws themselves are endowed with meaning. One could appeal to the integration of these laws in a still wider body of theory, but this recursive procedure is evidently of only limited help. Recursive procedures have to bottom out somewhere. The context account thus serves to explain how the meaning of particular scientific concepts is determined once the meaning of other scientific concepts is given; but it fails to elucidate the origin of meaning in the first place. We are still left with the question, what supplies us with the semantically laden point, on to which the context account could attach its machinery.

The answer is brief and simple: experience. It is experience that furnishes the semantic seeds that subsequently unfold through the theoretical integration. The crucial operation is the attribution of an intended domain of application to a law. This is exactly what we did in § 1 so as to set apart the substantively different interpretations of the same formal equation. The semantic ambiguity of a formal structure was removed by attaching referents to some of the symbols occurring in it. A definite meaning was conferred to the abstract variables by stipulating, say, that a particular variable was to refer to the elongation of a spring or, alternatively, to current intensity. Only if a domain of application is singled out in this fashion, can the network of laws do its useful job of fine-tuning the meaning of the concepts at hand.

Thus, we see that the meaning of scientific concepts cannot exhaustively be determined by the corresponding nomic framework. Or what amounts to the same thing, it can only be so determined if the laws involved are in advance semantically interpreted in the sense that their domain of application is fixed. We have thus to distinguish between two aspects of meaning. A semantic nucleus, as it were, that precedes the nomic integration of the concepts and originates from the attribution of referents to them in an at least loose and unsophisticated fashion, for one; and a theoretically generated meaning that derives from the concepts' figuring in a corpus of laws, for another. Reference must be given in a rough and intuitive way if meaning is supposed to be established by the corresponding network of laws. Let's look into this semantic share of labor taking General Relativity as an example.

Consider how a mathematical manifold can be interpreted physically so as to represent a space-time structure. Points in the manifold are interpreted as physical events; autoparallel, i.e., geodesic, null paths are construed as light rays and geodesic timelike paths as freely falling particle trajectories. Arc lengths of timelike trajectories are interpreted as proper time. Something of this sort must be given in order for us to deal with a physical theory at all – rather than a mathematical theory of (pseudo-)Riemannian manifolds.

Once a basic interpretation of the kind is available, the interpretation of other concepts can be clarified with the help of the theory. The affine connection or the Riemann curvature tensor, respectively, represent the gravitational field (cf. IV.3.§4, §8). Furthermore, a spacelike geodesic is demonstrably singled out through the criterion that the timelike geodesics starting off orthogonally from different points of a spacelike curve do not meet on either side (cf. Synge 1960, 155–156). Accordingly, relying on some interpretative input the theory manages to clarify the semantic import of a variety of concepts it contains. In this way it also contributes to a deeper understanding of the concepts used at the start. This is how the context theory of meaning is supposed to do its job.

In view of this consideration I wish to make the following two-fold claim. First, the attachment of referents, be it primary, be it derivative, must somewhere involve the distinction of measuring instruments. However, second, this distinction does not automatically include the capacity to conduct a full-blown theoretical analysis, along with the accommodation of distortions, of such an instrument. Consider time for example. If we want our algorithm to receive an empirical interpretation through the stipulation that arc length expresses itself as proper time, then this requires that proper time is at least crudely empirically accessible. And this in turn demands that at least one relevant measuring procedure is known in advance.

In other words, if an empirical interpretation is supposed to be imparted to a formal algorithm via the concept of proper time, then this requires that we be able to tell apart values of proper time from, say, the value of one's property. That is, we must be able to assess the adequacy of the measuring instruments we are presented with. We must be able to tell whether or not the data they furnish can justifiably be thought of representing a certain theoretical quantity. This shows that interpreting a theory empirically implies letting it operate as a criterion of adequacy with respect to the measurement of at least one quantity figuring in it (cf. I.3.§1, II.1.§2).

However, and this constitutes the second half of the above-made claim, this is not equivalent to letting it operate as its own observation theory. This becomes clear immediately by recalling an example referred to before repeatedly. We may, namely, accept the readings of an atomic clock as yielding the standard of proper time. Nothing prevents us from stipulating the appropriateness of atomic clock data as criterion of adequacy for proper time values. But as stressed time and again, the process furnishing nuclear time remains a black box from the perspective of General Relativity, that is, the theory introducing and specifying that quantity. In contradistinction to atomic clocks, however, General Relativity is perfectly capable of analyzing the mechanism, along with possible distortions, of light clocks such as the Marzke–Wheeler clock. This shows that the process of semantic interpretation of a theory leaves room for completing the theory; but it in no way guarantees the theory's completeness.

So, the overall conclusion is three-fold. (1) There is a semantic model in whose framework the notion of Einstein–Feigl completeness can be accommodated, namely, the theoretical context account of meaning. In its framework completeness becomes a sensible notion. Furthermore, in light of this model we obtain the following results regarding the process of semantic interpretation and its impact on completeness. (2) The theory in question must operate as criterion of adequacy with respect to at least one of its concepts in order to be able to operate as its own observation theory with respect to others. (3) Completeness is indeed a distinctive feature of a theory; it is not an automatic by-product of the process of giving a theory a semantic interpretation.

§ 3. After having considered the semantic aspects involved in completing a theory I now turn to the methodological ones. Again, I start by focusing, as a contrast, on constructive axiomatizations. It can be gleaned from Reichenbach's characterization, as sketched in V.3.§1, that a driving motive behind giving a theory a constructively axiomatized shape is to make the theory as directly testable as possible. The theory is to be grounded in elementary experiences that can be checked without relying on additional theoretical premises. Non-mediated testability is thus the goal.

Making a theory directly testable is to operate as a safeguard against two types of theoretical flaws. It is to make sure, first, that a theory stays in close touch with the phenomena it is supposed to cover and does not indulge in empirically useless theoretical ornaments. Second, dispensing with additional theoretical aids makes the validity check immune to errors possibly contained in them. It is clear, after all, that these auxiliaries may themselves be questionable or outright mistaken. The first item expresses a commitment to a sort of *empiricist principle of parsimony*: No empirically superfluous encasement should be erected around the facts; every theoretical element has to express an empirical phenomenon. The second item gives testimony to the conviction that a more directly tested theory is a better tested theory.

The chief motivation characterizing the completeness program is likewise a principle of parsimony though of a somewhat different nature.

A methodological reason for appreciating completeness is that it increases the *explanatory power* of a theory. A theory's explanatory power can be increased by enlarging the range of the phenomena it covers or by improving the accuracy of the theoretical treatment it gives or, finally, by reducing the number of independent theoretical premises it invokes. High explanatory power means that an extensive and precise empirical output is produced with the help of as little input as possible on the level of the fundamental assumptions.

It is immediately clear that a complete theory scores comparatively well according to this criterion. After all, the need of a theory to resort to auxiliary observation theories implies that the theory has to employ additional, independent laws so as to treat the relevant indicator processes. Incompleteness thus tends to diminish the explanatory power of a theory. Put conversely, relying on theory-dependent measuring procedures contributes to stepping up, *ceteris paribus*, that theory's explanatory power. That is, if completeness is not purchased at the price of a diminished empirical performance, it tends to increase explanatory power by way of a reduced theoretical input. A complete theory is thus supported to a higher degree than its incomplete counterpart in light of this methodological criterion.

It is clear that this criterion grows out of a principle of parsimony that directs us to minimize the number of theoretical assumptions considered as basic or elementary. This *theoretical principle of parsimony* thus leads in a different direction than its empiricist analog. Constructive axiomatizations usually operate with a vast number of axioms each of which expresses a supposedly elementary fact. There is a lot of elementary parts to be assembled in order to arrive at the sophisticated theoretical machinery. On the completeness view, every such axiom counts as a basic assumption with the result that the explanatory power of a so structured theory is evidently deplorably low. By contrast, the top assumptions of rather abstract formulations, as suggested by the completeness program, usually do fairly badly with regard to their direct testability. They hardly render elementary facts in the constructive sense. This contrast brings out clearly the difference in orientation between the two approaches. The difference is represented by the commitment to a theoretical or a specifically empiricist principle of parsimony, respectively.

§ 4. Finally, I wish to address the wider epistemological background of the completeness program. This background is represented by *coherentism*, i.e., by the view that the only epistemic virtue suitable for distinguishing a theoretical system is the coherence of the claims it embraces. Coherence requires, roughly and intuitively, that these claims are, first, consistent, and, second, mutually supporting (cf. Dancy 1985, 110–112). The second feature can be spelled out such that the corresponding system contains no elements that are isolated from, and thus unsupported by, the remaining parts of the theory. In other words, the theory should make maximal use of all conceptual elements it introduces. And this is in turn equivalent to a commitment to a theoretical principle of parsimony as explained in § 3. It will be seen shortly that the contribution of the completeness program to coherentism consists in providing a sort of feasibility proof. It helps to show that circularity problems threatening the viability of the coherentist approach can be tackled satisfactorily. In order to clarify the latter approach it seems advisable to pursue once more the track followed in the foregoing subsections. That is, I start by turning to the alternative approach, namely, foundationalism.

Foundationalism can roughly be characterized by the following two contentions.

(1) The *rock bottom postulate*: There is one epistemic basis of all scientific theories.

This is to say that all scientific theories eventually rest, however indirectly, on a common epistemic foundation that is not itself subject of any scientific theory. The relation between the foundation and these theories is drastically asymmetric in the sense that the former precedes and supports all epistemic claims rightly to be associated with the latter. Conversely, the former is not in need of any support from the latter; it rather stands on its own. Depending on the particular brand of foundationalism the epistemic foundation is identified with immediate sensory experiences, primary human skills or unquestionable facts of everyday life.

(2) The *hierarchy postulate*: Starting from this epistemic basis, scientific theories can uniquely and linearly be ordered by means of the two-place relation "is founded on". A theory T_2 is founded on a theory T_1 if T_2 makes use of concepts or methods developed within T_1.

The linearity claim is essential here. It is contended that the foundation relation allows for constructing a hierarchy among theories. It is precluded, that is, that T_2 import a certain concept from T_1 and at the same time export some other concept to T_1. The foundation relation between theories is as asymmetric as the one between the epistemic basis and a scientific theory.

It is thus claimed that all scientific theories finally rest on epistemically distinguished entities whose distinction derives from the fact that no theories (i.e., no observation theories) are needed for their identification (the rock bottom postulate). The rock bottom basis is then supposed to allow for a stepwise construction, without gaps or circles, of scientific quantities proper. The emerging structure of science is represented by a hierarchy of theoretical layers ordered by the foundation relation (the hierarchy postulate). It appears now that constructive axiomatizations are inspired by the empiricist brand of foundationalism. As an example of the foundation relation think of the concept of temperature as specified within phenomenological thermodynamics. This concept is then exported to statistical mechanics where it is identified with the mean kinetic energy per degree of freedom. Statistical mechanics is thus interpreted through, meaning: is founded on, phenomenological thermodynamics.

§ 5. Foundationalism has a bad press nowadays, and rightly so. First, the task of identifying the postulated rock bottom elements, theoretically disinfected in themselves but capable of supporting the epistemic claims associated with science, has turned out to be fairly hard, to put it mildly. Second, the hierarchy view has given way to a less asymmetrically structured picture of mutual support and explanation. Consider the example of the *Grand Geometric Circle*. We start with an everyday version of physical geometry, i.e., with a Euclidean-type theory of paths and length relations. If basic methods for measuring durations are specified in addition, we are able relying on this primary theory to evaluate spatiotemporal quantities. On this basis the second floor of classical kinematics is erected. This theory gives a description of motion without considering its possible causes. The latter are addressed on the third floor, namely, classical mechanics. Now, forces are taken into account and particular force laws, such as the law of gravitation, are specified. Next follows the fourth floor of classical electrodynamics. The new property of charge is taken into consideration and the pertinent laws, i.e., Maxwell's equations and Lorentz's force law, are added. On the fifth floor, finally, General Relativity makes its appearance. The theory unifies and integrates classical mechanics and electrodynamics – which is still in keeping with the foundationalist picture; but at the same time it leads us right back into the first floor – which squares pretty bad with it. Through this final bend the supposed hierarchical structure is transformed into a Grand Circle in which the higher-level entities, namely, gravitational influences, act back on the lower-level ones, namely, geometric relations. What happens on the highest floor modifies the structure of the foundations which consequently cease to be foundations anymore.

It appears immediately that bidirectional influences of this sort are grist to the coherentist mill. Reciprocal clarification, rather than unidirectional foundation, is the key relation characterizing the structure of science. Coherentism stresses the reciprocal elucidation of theory and evidence guided by the overall maxim that the resulting body of laws be maximally coherent. This implies a preference for the theoretical clarification of measuring procedures, and it is here where coherentism meets the completeness program.

Now, the point is that reciprocal clarification naturally leads into loops and circles. If the coherentist approach is supposed to be viable it has to be shown that these loops and circles are not vicious in character, i.e., that they can be handled. And this is precisely what the present elaboration of the notion of completeness is intended to demonstrate. This is why the successful realization of the completeness program contributes to showing the feasibility of the coherentist approach. Since coherentism is committed to a theoretical principle of parsimony (see § 3), the completeness program thus helps to clarify how a tough conceptual austerity program can be implemented.

5. CONCLUSION

Once more it is time to order our thoughts and to bring out more succinctly the various aspects developed in this chapter. The question guiding the investigation is whether General Relativity is Einstein–Feigl complete regarding measurements of spatiotemporal quantities; and if so, whether this feature also extends to indicator attachments under non-ideal circumstances. Put more concretely, we are asking whether it is possible relying on the theoretical apparatus of General Relativity to specify measuring procedures suitable for the evaluation of the basic quantities of physical geometry, namely, geodesic paths, for one, and length and duration relations, for another. And moreover, whether it is possible to analyze the impact of factors distorting the measurement and to sift out the pure states from perturbed data. So, our project is to clarify what it takes to translate the abstract geometric machinery into concrete evidence.

An important step in this direction can be made using the Ehlers–Pirani–Schild approach to constructing physical geometry from easily accessible facts such as particle trajectories and light rays. Relying on basic facts of this sort a criterion can be specified which allows for singling out timelike geodesic motions. This is equivalent to the capacity of empirically setting apart freely falling test particles from other particle motions. That is, if we have agreed to consider test particle motions in the exclusive presence of a gravitational field as perturbation-free, then the Ehlers–Pirani–Schild construction puts us in a position empirically to identify the so characterized pure states of motion without drawing heavily on additional observation theories and without invoking the higher-level machinery of General Relativity, i.e., the field equations (cf. V.1.§2–3).

Furthermore, it is possible to construct on this fairly narrow conceptual basis instruments suitable for detecting the prevailing metric relations. Such instruments are represented by light clocks which indeed make exclusive use of test particle paths and light rays. An example is the Marzke–Wheeler clock which basically consists of a light pulse reflected back and forth between two parallel particle trajectories. The task is, then, to specify a parallel to a given geodesic trajectory without resorting to metric relations. And this is precisely what the Marzke–Wheeler construction achieves. Relying on the Marzke–Wheeler clock, arbitrary intervals can be evaluated metrically (cf. V.1.§4).

However, there is still an important proviso that has to be removed if this conclusion is to be sustained. Namely, we have to secure the reliability of the clock in the possible presence of distortions. What is of particular relevance here is the impact of geometric distortions. Such distortions originate from curvature influences. The presence of non-vanishing curvature may result in the deformation of extended bodies or in the deviation of their trajectories from the corresponding geodesic courses. Moreover, the parallelism of particle trajectories is destroyed under these circumstances. It is the latter effect that is detrimental to the proper operation of the Marzke–Wheeler clock. The

Marzke–Wheeler construction only works appropriately in the absence of curvature; otherwise, the two trajectories that constitute the clock are pushed off their parallel tracks through the action of tidal forces.

It turns out that the influence of curvature on the correct functioning of the Marzke–Wheeler clock bears the marks of a self-referential distortion. The distorting effect of curvature on the proper-time readings supplied by the clock can only be evaluated and consequently rectified if reliable proper-time readings are already at hand. This self-referential relation springs from a Reichenbach loop connecting the empirical determination of curvature and proper-time intervals (cf. V.1.§5).

However, this corrective circle can be defused by resorting to the standard procedure already familiar from previous chapters, namely, the realization of the pure case. It is indeed possible to approach the undistorted state by performing a limiting sequence of clock experiments. This is accomplished by a continual reduction of the clock's period of operation until the evaluation of a given interval, as provided by the clock, no longer changes on further reduction (cf. V.1.§6). This scenario is conceptually analogous to the establishment of a unique relation of distant simultaneity by means of infinitely slow clock transport (cf. IV.5.§2). The pure case is identified but turns out to be empirically inaccessible. So, an iterative procedure is specified that brings us ever closer to the distinguished state. The infinite regress generated by a corrective Reichenbach loop is stopped by advancing toward the pure state in which the relevant perturbation has ceased to be influential.

Furthermore, the identification of the pure state proceeds through external fixing. The pure state is represented by a shortly operated freely falling Marzke–Wheeler clock. Under these circumstances curvature does not become manifest. This fortunate trait is guaranteed by the properties of Riemannian space-time; and it is the assumption that such a space-time structure is actually realized that serves to fix externally one of the quantities making up the Reichenbach loop under discussion. As just mentioned, this loop is generated by a reciprocal dependence between the evaluation of curvature and proper-time intervals, so that the control of curvature influences by way of external fixing make time intervals measurable. Moreover, what appears striking is that a different consequence of the same assumption of Riemannian space-time being realized is instrumental in analyzing the diverging influence of curvature on particle trajectories and thus helps to shed light upon the nature of the distorting factor at hand. We are thus presented with an especially economic brand of external fixing (cf. V.1.§7).

This result leads to the conclusion that by exclusive recourse to General Relativity we are able to devise an instrument suitable for measuring spatiotemporal metric relations; and that we are likewise able to analyze the impact of geometric distortions on its operation and to specify procedures that allow us to discount cases spoiled this way. Hence, we can get hold of the pure states even if (particular types of) distortions are present. General Relativity is thus sufficient for attaching empirical indicators to its geometric notions under ideal conditions as well as under geometrically distorted ones. And this is to be expressed so that General Relativity is Einstein–Feigl complete with respect to the evaluation of spatiotemporal quantities (cf. V.1.§7; V.3.§3).

We are not done though, for there is one more safeguard operation necessary. This one concerns the perturbing impact of eventual differential distortions, i.e., distortions whose intensity varies with the particular nature of the material used for carrying out measurements. The point is that differential distortions are likewise self-referential in character. In order to correct, e.g., the influence of thermal expansion on rod-based measurements, a reliable standard of rigidity must be available beforehand. Every trustworthy determination of metric relations by means of rods thus demands that dependable metric data be at hand in advance. Accordingly, we are faced with a reciprocal dependence between differential distortions, on the one hand, and the resulting metric, on the other. And this relation appears to be precisely analogous to the corresponding dependence between Reichenbach's universal forces and the metric. In other words, this scenario seems to constitute a differential analog to Reichenbach conventionality (cf. V.2.§1).

Fortunately enough, this is not quite true. For the distorting influence and the undistorted quantity we are after are not related by merely one single hypothesis but by a larger number of laws. This is why we are dealing with a Grünbaum helix rather than a Reichenbach loop. A Grünbaum helix allows for specifying a procedure of reciprocal adaptation that finally provides us with a distinguished value for either quantity (cf. V.2.§3). Since, on the other hand, the situation is complicated by some technical intricacies in the thermal scenario under discussion, this helical procedure is of no general avail for defusing the pertinent self-referential distortion circularity (cf. V.2.§4). Still, a Grünbaum helix in principle represents an additional option for performing corrections in complete theories.

In addition to the just-described procedure, there exist three further options for tackling the differential correction circularity problem. What we just discussed represents an attempt to correct distorting influences in rod-based measurements. But we might as well try to select undistorted instances rather than correct the distorted ones. Furthermore, the same pair of options exists with respect to measurements performed on the basis of particle trajectories and light rays. In this framework differential distortions are represented, e.g., by a lack of neutrality of the moved particles. Charged particles in an electromagnetic field deviate from a geodesic path. The point is that this scenario likewise gives rise to a self-referential circularity that is represented by a reciprocal dependence between the metric to be determined and the correcting Lorentz force.

As it turns out, neither the selection option if applied to rod-based measurements nor the correction option if put in operation with trajectory-based ones is of any help (cf. V.2.§5–6). This being the case, the last resort scheme is to bring to bear the selection option to trajectory-based measurements. Mercifully, we can take advantage of the Ehlers–Pirani–Schild criterion for geodesic motion so as to single out differentially undistorted trajectories. Application of this criterion does not involve any metric relations so that the pure states of motion can indeed be identified without prior recourse to data that are only provided by reliable measurements. The corrective circle is thus broken up by singling out cases that are not in need of any correction (cf. V.2.§7).

The essential precondition for employing the selection option is that there is no argument about what qualifies as the pure state. And other than in the universal-force scenario (cf. IV.4.§4) or the distant-simultaneity scenario (cf. IV.5.§6) agreement about the pure state is easily reached here. It is represented by the motions of test particles in a gravitational field. It is clear, on the other hand, that even if the nature of the pure state is undisputed, such states can only be selected if they exist in fact. And their existence obviously constitutes a contingent feature and is not a matter of course. Accordingly, the differential correction problem is not resolved in full generality. But since as a matter of fact the pure states are given in experience, the above-stated conclusion as to the completeness of General Relativity can indeed be maintained. It is possible to attach indicators in a self-sufficient fashion to the fundamental geometric notions of General Relativity. This is accomplished by separating the wheat from the chaff, i.e., by selecting instances which correspond to these fundamental notions.

The next question is whether the General Relativity's capacity to connect indicator phenomena to theoretical states also extends to distorted situations. Up to now we have only considered its capacity to differentiate by its own lights between pure empirical states and distorted ones and to license the inference from the former to the corresponding theoretical states. So, we now ask whether General Relativity is sufficient also to establish a connection between distorted empirical states and theoretical notions. Can we do more with the distorted cases than simply do away with them?

We can indeed. By bringing to bear the full-scale field equations we are able to extract the prevailing metric relations from distorted trajectories. General Relativity thus allows for specifying a correction process that makes it possible to infer theoretically relevant quantities from disturbed evidence. Relying on the theory we are able to assess the theoretical bearing of perturbed instances (cf. V.3.§4). The result is, then, that General Relativity is complete with respect to attaching indicators and to licensing rectification procedures. In other words, it justifies the relevant state-indicator connections under ideal and non-ideal conditions alike. In particular, General Relativity is sufficient but not necessary for both purposes (cf. V.3.§4).

The final issue addressed concerns the clarification of the wider semantic and methodological background of Einstein–Feigl completeness. I tried to shed some additional light on how completeness is possible and why it is desirable. What appears to threaten the mere possibility of completing a theory in the first place is that no abstract formalism can determine the domain its variables are supposed to range over. Putting such a formalism to empirical work always demands a stipulation as to the empirical interpretation of at least one of the corresponding variables. This stipulation has to be appended to the formalism and cannot be derived from it. If completeness is supposed to accomplish such a derivation, no theory can ever be completed and the whole project only gives testimony to a great deal of semantic confusion (cf. V.4.§1).

Hence, completeness has to be conceived in a different fashion. If it is to constitute a sensible notion it has to refer to theories that are already empirically interpreted. The theoretical context account of meaning provides the semantic framework appropriate for this task. According to this model, the meaning of a theoretical concept is determined by the network of empirically interpreted laws in which it figures. This being the case, we are left with the challenge of elucidating the origin of the meaning of such concepts. Regarding scientific concepts, there is clearly only one option for doing this, namely, by drawing on experience. Empirically accessible phenomena have to be associated with some of the concepts under consideration. Only if a theory's intended domain of application is fixed this way, can the corresponding nomic relations contribute to determining the empirical interpretation of the remaining concepts and to deepening and ramifying the semantic import of those interpreted initially (cf. V.4.§2).

But now we appear to be faced with a problem that is quite the reverse of the one we were tackling before. For the result is that every empirical theory in virtue of being an empirical theory exhibits connections to relevant indicator processes. Every such theory is essentially tied up with indicator processes; if not, it is not an empirical theory. Accordingly, completeness appears to be achieved automatically. Rather than an impossibility, Einstein–Feigl completeness seems to represent a mere triviality.

Fortunately enough, however, this is not the case. For while it is true that the process of interpreting a theory empirically involves at some stage or other the distinction of certain measuring instruments as appropriate means for evaluating some of the theoretical quantities at hand, this distinction is not equivalent to the capacity to perform a theoretical analysis of these instruments. The disparity between distinguishing an instrument and analyzing its mode of operation is rendered by the difference between using a theory as criterion of adequacy and using it as observation theory. Employing a theory in the former sense does not entail the capability of employing it in the latter sense as well. The process of interpreting a theory semantically does not automatically bring that theory's completeness in its train. For this reason completeness indeed represents a distinctive feature of theories; and, consequently, it is definitely a non-trivial task to make a theory Einstein–Feigl complete (cf. V.4.§2).

The final question is why we should bother at all about embarking on such an endeavor. What makes completeness desirable and worthwhile to pursue? The chief distinction of complete theories is their high explanatory power. In complete theories the corresponding empirical data can be processed without recourse to further, independent observation theories. It is the theory itself that is sufficient for analyzing the operation of the relevant instruments and for correcting the raw-values they supply. The basic principles of complete theories thus do a lot more explanatory work than do the assumptions heading incomplete ones. There is more explanatory delivery for a given theoretical cost. Accordingly, a complete theory excels in theoretical parsimony. In such a theory the number of independent fundamental assumptions is kept at a minimum. This is what makes completeness an appealing feature (cf. V.4.§3).

At least this holds if completeness is not acquired in exchange for some unpleasant shortcomings. A vice in point would be represented by reduced testability. If the testability of a complete theory is diminished as compared to its incomplete counterpart, then we better forget about completeness altogether. But nothing of the kind turned up in connection with General Relativity Theory, the reason being that it was completeness on the theory level, rather than on the hypothesis level, that was of primary concern in this chapter. The completeness of General Relativity is thus in no way epistemically suspicious.

What is noteworthy here, in addition, is the high degree of completeness this theory enjoys. Moreover, General Relativity readily and naturally lends itself to an interpretation in terms of completeness; this interpretation is not forced on the theory. This marks a striking contrast to quantum mechanics which is generally thought to represent the most fundamental and most comprehensive physical theory. It will be recalled from III.2 that quantum mechanics has to struggle with completeness. More specifically, quantum mechanics can only be completed, given the present state of knowledge, by resorting to appalling, rather than appealing, expedients.

I do not intend to suggest, however, to take this marked ranking difference on the completeness scale as an argument for supplanting quantum mechanics with an approach fashioned after the model of classical physics. First, completeness does certainly not constitute *the* preeminent methodological quality that could eclipse all other methodological virtues. Second, it is to be expected that sooner or later quantum mechanics will be supplemented with a working quantum theory of measurement. Third, if not, the fault may rather lie with nature than with man. Maybe the incompleteness of quantum mechanics has to be swallowed as a consequence of the facts of nature; there may exist no sensible way to complete quantum mechanics. We have to bear in mind that the structure of scientific theories is not entirely at our disposition. Nature does a lot to impose severe constraints on them. So it may well turn out that the facts prevent us from completing quantum mechanics in a reasonable fashion. Classical physics cannot be recaptured by way of methodological postulates.

CHAPTER VI

THE CONVENTIONALITY OF PHYSICAL GEOMETRY: A RECONSIDERATION

A lot of symphonic movements end up with a *coda*. The apparent reason is not to push the audience too abruptly back into the hardships of ordinary life. There even exist cases in which the *coda* is a dramatically important part of the corresponding movement and contains new thematic material. Consider the Second Movement of Bruckner's Seventh Symphony or Strauss' "Death and Transfiguration". True, a *coda* rather marks the end of a movement and consequently does not constitute a movement of its own. But as we all are well aware, there is no rule of art that has not yet been breached. So, I feel free to indicate that this chapter represents my *coda* movement.

The issue I wish to address is the conventionality of physical geometry. I have already referred to this issue repeatedly. I described the argument adduced in favor of the conventionality claim (see IV.2 and IV.4), and I discussed the underlying difficulty in terms of completeness (see IV.4 and IV.5). In the latter context, the conventionality problem was intended to instantiate and highlight the epistemological trouble one may run into while completing a theory. What I want to do now is to analyze the viability of the conventionality claim proper. The question is whether or not physical geometry is conventional in a philosophically interesting sense. So, the intent is again systematic but it differs somewhat from the more methodological objective of the earlier discussion. To state it right at the beginning, the answer will turn out to be affirmative. To wit, geometry is conventional in important respects.

As described in IV.2 and IV.4, the conventionality thesis says that experience leaves room for several, different geometric theories. It is possible to give conceptually distinct but nonetheless empirically equivalent accounts of observed metric relations. Accordingly, every choice from among the class of empirically equivalent alternatives has to proceed by invoking non-empirical criteria. And this leads to the contention that adopting a particular version of physical geometry always contains a conventional element.

In the course of the last three decades, geometric conventionalism has virtually been put out of business. In my view this is not due to its poor competitiveness. True, the balance of conventionalism has been written in red ink for years but this only indicates the need for an overhaul. So, I intend to put conventionalism back on an even keel. In particular, I am going to argue that the conventionality thesis is justified to an even greater extent than its classical advocates had foreseen. In its traditional form the thesis is confined to the empirical consequences of conceptually different approaches. This means, roughly speaking, that the empirical equivalence of two accounts gives us sufficient reason for labeling them conventional alternatives – although they may fare vastly different in methodological respects. I think, however, that the conventionalist position can and should be strengthened. Even if one takes methodological criteria into account, there remains a conventional element in the choice among empirically equivalent alternatives in physical geometry. This conventional element stems from the conflict between different criteria of this sort.

In defending this contention, I will proceed along the following lines. In the first section I elaborate the just-mentioned stronger concept of conventionality, and in the second one I apply this concept to two theories of physical geometry, namely, the Newtonian theory of gravitation and General Relativity Theory. My claim is that in both cases alternatives can be identified that qualify as conventional alternatives in the stronger sense. So, let's set out to get conventionalism off the hook.

1. POINCARÉ SET STRAIGHT: HIGH-PROFILE CONVENTIONALISM

§ 1. It will be useful briefly to rehearse the traditional conventionality argument as it originates with Poincaré and is given its mature expression in Reichenbach. Reliable length measurements require that the rods employed are rigid, i.e., that they are not subject to deforming influences. But consider a Reichenbachian universal force that acts alike on rods of all materials. Such a force tends to deform all measuring rods in an equal fashion, and for that reason its presence or intensity is hard to detect. This does not mean, however, that it escapes all detection. If the universal force field is inhomogeneous, i.e., if its strength is different at different locations, then it induces empirically accessible deformations in rods that are sufficiently long to extend over these inhomogeneities. Moreover, because the counteracting cohesive forces in the rod's material are substance-specific, the deformations induced by the universal force may vary with the material employed (cf. IV.4.§1–2).

The conventionalist point here is that we have two different options for coping with such a situation. On the one hand, we may treat such a force as a distortion and its effect as an alteration of the rod's length. On this approach a rod placed in a universal force field is considered deformed and thus ceases to correctly indicate the prevailing geometric relations. The presence of such a field destroys the rod's rigidity. Regarding the spatial metric g_{ab} $(a,\ b = 1,\ 2,\ 3)$ this means that the "true" metric ${}^*g_{ab}$ is veiled by a universal influence U_{ab} with the result that g_{ab}, distinct from ${}^*g_{ab}$, turns up empirically:[1]

(1) $${}^*g_{ab} + U_{ab} = g_{ab}$$

Consequently, in order to obtain reliable indications of the prevailing geometry, the intensity of the distorting field U_{ab} has to be evaluated and its deforming effects have to be corrected.

On the other hand, we may geometrize away a universal force. On this interpretation the rod's deformation is not construed as arising from a force but as an effect of space-time curvature instead. The deformation indicates changes in the metric structure of space-time. Accordingly, the deformation is not to be viewed as a distortion; rather, it truthfully indicates actual differences in the geometric structure under scrutiny. Hence it makes no sense to correct it, and this means that a rod placed in a universal force field retains its rigidity.

The two approaches are empirically equivalent. The observed coincidence relations among measuring rods can equally be accommodated either way. Accordingly, as conventionalism contends, it is not a fact of the matter which of them is true. Rather, it is up to us to make a decision among them, and we make it on grounds of simplicity and convenience, that is, conventionally. On the basis of such criteria it appears preferable, as Reichenbach argues, to dispense with universal forces and to adopt a non-flat geometry. However, what lies at the heart of the conventionalist position is the claim that considerations of simplicity and convenience cannot confer epistemic superiority to a theory. The latter can only be based on that theory's empirical record. As regards epistemic merit, all empirically equivalent theories are on the same footing (cf. Reichenbach 1928, 30–37).

Moreover, as explained in IV.4.§5, the argument can be given a spatiotemporal shape by referring to particle motions rather than rod-based measurements. In that framework, universal forces are introduced, or removed as the case may be, by molding the law of inertia. Suppose we encounter a particle that is moved inertially with respect to a given affine connection ${}^*\Gamma^i_{jk}$. We may now easily introduce a universal field U^i acting independently of any particle properties. On that account, the same particle motion comes out as accelerated with regard to a changed connection Γ^i_{jk}. Its equation of motion (that was already given as Equation (6) in IV.4.§5) is then:

$$(2) \qquad \frac{\mathrm{d}^2 x^i}{\mathrm{d}\tau^2} + \Gamma^i_{jk} \frac{\mathrm{d}x^j}{\mathrm{d}\tau} \frac{\mathrm{d}x^k}{\mathrm{d}\tau} = U^i ,$$

The universal field can be further specified by subtracting the equation of geodesic motion from Equation (2). The result is:[2]

$$(3) \qquad (\Gamma^i_{jk} -^* \Gamma^i_{jk}) \frac{\mathrm{d}x^j}{\mathrm{d}\tau} \frac{\mathrm{d}x^k}{\mathrm{d}\tau} = U^i ,$$

This maneuver amounts to reinterpreting geodesic motion as force-induced motion with respect to a changed geometry. What was viewed as motion along the straightest line possible before is subsequently construed as a curved trajectory.[3] Provided that light rays are thought to be affected likewise by these universal influences, the geometrized account and its universal-force analog lead to the same observational consequences. Accordingly, the concept of inertial motion is no more uniquely specified empirically than the concept of rigidity is. In both cases the facts leave room for alternative approaches so that a preferred account can only be singled out by recourse to non-empirical criteria.

§ 2. The conventionalist position has come under heavy fire by realistically minded philosophers. According to scientific realism, the central concepts of our most successful theories refer to theory-external, i.e., really existing, entities. One realizes immediately that conventionalism and realism are at odds with one another. A necessary precondition for the viability of scientific realism is that there are no serious or non-trivial conventional elements involved in scientific theories. In other words, if science is supposed to unveil the real nature of things, the particulars of scientific accounts must not depend on the idiosyncratic factors of human investigators.[4] Conversely, the tenability of a conventionalist position tends to support a non-realist interpretation of science.

Realists contend, accordingly, that the metric field is not only a convenient shorthand-description of measured length values and acceleration relations. Rather, it represents a physical field in its own right (cf. Putnam 1963, 94–95). Consequently, realists claim that there is merely a trivial semantic brand of conventionality involved in specifying the metric. We could have employed the word "metric" to denote something different (cf. Putnam 1975, 167). But from the epistemological point of view the metric is quite definite; it is provided in an unambiguous fashion by General Relativity Theory. In this vein Friedman asserts:

The space-time of general relativity is endowed with a perfectly definite metric [$ds^2 = g_{ik}\, dx^i\, dx^k$], which is related in a definite way to the distribution of mass-energy by Einstein's field equations. There is no sense in which this metric is determined by arbitrary choice or convention (Friedman 1983, 26).

This quote renders the definiteness thesis which has now gained predominance over the rivaling conventionality thesis within the philosophical community.

Underlying the definiteness thesis is, first, the theoretical context account of meaning. On that account, a concept's meaning is fixed through the theoretical network it forms part of (cf. I.2.§3). The concept's reference is derived from its meaning and is thus likewise determined in this fashion. This means conversely that reference is not determined through the simple attachment of an operational criterion. It is fixed not by simply stipulating a coordinative definition; rather, the relevant theoretical postulates are drawn upon for this purpose as well.

However, this move was already anticipated by conventionalists. It is true, as Reichenbach argued, that it is sometimes possible to specify a concept's reference by appeal to its theoretical integration. We may, for instance, employ the law of inertia so as to determine equal periods of time. Yet the catch is that we may as well tamper with the law of inertia and consequently arrive at a different judgment about what is to count as equal durations (cf. Reichenbach 1928, 116). Emphasis on theoretical integration thus merely shifts the conventionalist free play from the operational definitions to the corresponding corpus of laws.[5] The conventionalist claim is, then, that experience is sometimes unable to single out a definite corpus of laws.

In fact, this is admitted implicitly by scientific realists for there is a second step adduced to the anti-conventionalist response. It is claimed that we should not consider all empirically equivalent alternatives equally suitable for determining reference. Rather, we assess their suitability for this purpose in light of methodological criteria. Consider Hilary Putnam's version of this strategy. We appraise the adequacy of theories by means of two sorts of criteria, namely, agreement with experience, for one, and simplicity and agreement with intuition, for another. Truly conventional alternatives have to score equal on both scales. A universal-force account of physical geometry, by contrast, is "*improbably*" complicated and counter-intuitive and is thus to be considered false. There is no conventionality of physical geometry besides trivial semantic conventionality (cf. Putnam 1975, 165–167).

This argument obviously hinges on the proposition that a more complicated but empirically equivalent account is "improbable" in the sense of being less likely to be true. Put more generally, in order for the argument to be persuasive a connection between methodological qualification and truth has to be established. But this missing link is hard to supply – at least with respect to the methodological criteria advanced. Why should a simpler theory, other things being equal, portray reality more faithfully? Hardly anybody would still be willing to resort to the inherent simplicity of Nature herself to buttress such a contention. As Duhem put it, the physicist

> will choose a certain formula because it is simpler than the others; the weakness of our minds constrains us to attach great importance to considerations of this sort. There was a time when physicists supposed the intelligence of the Creator to be tainted with the same debility, when the simplicity of these laws of nature was imposed as an undisputable dogma [...] That time no longer exists. We are no longer dupes of the charm which simple formulas exert on us; we no longer take that charm as the evidence of a greater certainty.[6]

Well, it seems as if Duhem's conclusion as to the actual behavior of physicists, or for that matter philosophers of science, had been a bit premature. Still, it is hardly possible to convincingly ground the distinction of simple and intuitively appealing theories on ontological considerations. I conclude that conventionalism should not be rejected off-hand and deserves closer scrutiny.

§ 3. Let us look into what an interesting and possibly tenable conventionalist position could be like. In other words, what precisely is the claim the arguments sketched in § 1 are suited to vindicate? What can the contention that there exist conventional alternatives in physical geometry sensibly mean? In the remainder of this section I try to capture the gist of that contention by specifying four requirements for alternative approaches that are to count as conventional alternatives.

The first condition is *empirical equivalence*. This demand is obvious and undisputed. For clarity's sake I hasten to add that equivalence really means equivalence here. That is, a mere approximate agreement of the observational consequences is not sufficient. Empirical equivalence further implies equality of scope. If one theory has a larger domain of application than another one while agreeing observationally with the latter within their common intersection, the two are not empirically equivalent.

§ 4. The second condition is the *non-synonymy* of the respective accounts. This is intended to exclude cases in which both accounts are merely different formulations of one and the same theory. A meaningful conventionality claim should discount such cases. The assertion that the choice from among different formulations of the same theory has no factual basis and is just a matter of convenience constitutes a truism rather than an interesting philosophical claim. The same story can always be told in different words. Right, but so what?[7]

The condition of non-synonymy can be expressed by the requirement of *non-intertranslatability* of the theoretical terms involved on both sides. This requirement is admittedly at variance with the conventionalist tradition. For instance, one of Poincaré's arguments for the conventionality of geometry is the intertranslatability of geometric theorems (cf. Poincaré 1902, 69). Moreover, he explicitly compares this conventionality to the conventional choice between different units and between Cartesian or polar coordinates (cf. Poincaré 1902, 76). Here, the alleged conventional alternatives are intertranslatable as well (cf. IV.2.§2, IV.4.§1).

But if that's all what is at stake the whole issue is a bore. Unit choices are evidently conventional, and the same goes for the choice of coordinate systems. Space-time theories are formulated in a generally covariant fashion, i.e., the relevant equations hold in (almost) arbitrary coordinate systems. But the coordinate representations of the quantities involved may look different according to the coordinate system chosen. The same metric will assume a different shape when being written in (locally) Cartesian or polar coordinates. If there is nothing more involved here let's forget about the conventionality of geometry. In addition, space-time theories can be rendered in a coordinate notation (as I did here all along) or in a coordinate-free representation of geometric objects. There is a straightforward correspondence between formulas in both renderings; and there is no interesting sense in which these two formulations could count as alternatives. Everybody may use either one as she pleases. So if that's supposed to be an instance of the conventionality of physical geometry, I suggest stop wasting ink on the subject.

Accordingly, in order to put some teeth into conventionalism we should demand non-intertranslatability on the theoretical level. The notion of intertranslatability can be clarified by invoking the concept of *Tarski-interpretability*. A theory T_1 is Tarski-interpretable in a theory T_2 if to each descriptive predicate of T_1 an expression of T_2-predicates can be assigned such that the T_1-theorems are preserved. In that event, the T_1-predicates are definable in terms of T_2-predicates. These definitions are translation rules that establish a connection or map between the respective theoretical vocabularies. The criterion of adequacy for these definitions is that all theorems of T_1 are mapped into theorems of T_2. Tarski-interpretability of T_1 in T_2 means, consequently, that all T_1-predicates can be translated into T_2-predicates such that all T_1-theorems become translated T_2-theorems.[8]

Interpretability is not a symmetric relation for it is not required that there be a one-one correspondence between the pertinent predicates. One T_1-predicate may as well be mapped onto a logical function of T_2-predicates. This implies that the definitions of T_1-predicates in terms of T_2-concepts do not necessarily constitute definitions of the T_2-concepts in terms of the T_1-predicates. But since synonymy is a symmetric concept we clearly have to add something. We should demand *mutual interpretability*, i.e., T_1 is to be interpretable in T_2 and, conversely, T_2 in T_1. It is clear that the set of definitions used in either direction need not coincide. That is, mutual interpretability is possible without one-one correspondence of the predicates involved.[9]

An important property of mutual interpretability is the preservation of the non-theoretical status of statements. Mutual interpretability preserves non-theoremhood. Unilateral interpretability of T_1 in T_2 is compatible with T_2's having a wider scope than T_1. That is, whereas all T_1-theorems become T_2-theorems upon translation, there may exist T_1-non-theorems whose translations indeed are T_2-theorems. Mutual interpretability excludes this possibility. For if the situation is as described, the attempt to interpret T_2 in T_1 would involve a map of theorems into non-theorems; and this vitiates interpretation. Mutual interpretability thus demands that theorems go into theorems and non-theorems into non-theorems.

Mutual interpretability is advanced here as criterion for distinguishing between different versions of the same theory, on the one hand, and different theories, on the other. The former are mutually interpretable and the latter are not. Since I wish to remove the former from the domain of application of the conventionalist claim, conventional alternatives are supposed to be distinguished by their empirical equivalence and their lack of mutual interpretability.

The conventionality of distant simultaneity, as described in IV.4.§6 and IV.5.§7, represents a contentious issue in the conventionality debate. So let's ask whether deviant synchronization procedures lead to conventional alternatives to standard Special Relativity if judged by the present non-synonymy criterion. The problem thus is whether the standard version, as characterized by isotropic light propagation and $\varepsilon = \frac{1}{2}$, and non-standard alternatives are mutually interpretable. In fact, the problem turns out to be rather tricky in that the answer depends on what precisely the standard version is supposed to be like. Usually, Special Relativity is presented using inertial frames of reference with straight coordinate axes. Non-standard versions are then characterized by the assumption of a direction dependence (or even position dependence) of the one-way velocity of light with respect to such a reference frame. Provided that constraints such as unchanged round-trip velocity are satisfied and that the synchronization procedure is suitably adapted to the one-way velocity chosen, the assumed direction dependence has no observational impact.

In that case the standard and non-standard quantities are related to one another as follows. The relevant equation is $c = 2\varepsilon c_{ow}$, with c representing the standard or two-way velocity of light and c_{ow} its one-way velocity in the positive x-direction.[10] It is realized that this equation provides a definition of the standard quantity c in terms of any given non-standard ε and non-standard c_{ow}. The reverse does not hold though. If we are only given standard quantities, i.e., c and $\varepsilon = \frac{1}{2}$, no non-standard quantity can be obtained. Plugging in $\varepsilon = \frac{1}{2}$ makes the equation collapse into the standard relation $c = c_{ow}$. The conclusion is that the standard version is interpretable in the non-standard version, but not the other way around so that mutual interpretability does not hold.

A different conclusion emerges, however, if restriction to inertial frames is dropped. As already indicated in IV.5.§7, Special Relativity can be given a generally covariant formulation; that is, it can be present-

ed using curvilinear coordinate axes and arbitrarily accelerated frames of reference. Standard and non-standard versions can then be identified with particular coordinate representations of the fundamental generally covariant equations. And these various coordinate representations are easily translated into one another by using coordinate transformations as translation rules (cf. Havas 1987, 439–441). On this more comprehensive approach intertranslatability obtains, and the different versions come out as different formulations of the same theory.

§ 5. Still, even if the non-synonymy condition is accepted conventionalism does not entirely lose its truistic ring. For it is clear that one can always introduce some superfluous complications into a theory; complications, that is, which serve no theoretical purpose and have no empirical effect. This can be done, for instance, by inventing some wild theoretical mechanism whose effects are compensated by a subsequent, equally wild mechanism. That is, theories can always be "deoccamized" at will (cf. II.4.§1). Deoccamization is not ruled out by the non-synonymy condition. For whereas the more parsimonious account can be interpreted in its lavish analog, the reverse is not true. The laws governing the functioning of the deoccamizing mechanisms obviously have no counterpart in the more pedestrian variant. This means, on the whole, that our two criteria are satisfied by pairs of theories that intuitively would not qualify as even remotely sensible alternatives.

In light of these considerations it seems appropriate to strengthen the concept of a conventional alternative by adding two further demands. These demands are intended to express a minimum standard of reasonableness. We should require *preservation of causality*, for one, and *non-vanishing methodological rating*, for another. Let's now turn to the discussion of these items.

Preservation of causality demands that no uncaused events occur. More precisely, if there are two empirically equivalent accounts one of which allows for attributing every relevant phenomenon to a cause according to a general law and the other one does not, then the causally regular account is to be preferred. This consideration bears on the case of physical geometry as follows. The conventionality thesis is open to the following objection. The whole argument, as rehearsed in § 1, is confined to the kinematical side of the situation and leaves its dynamical aspects untouched. More specifically, there was a lot of talk about induced deformations of rods and induced deviations from geodesic motions; but comparatively little has been said as to the possible *causes* of these effects. The universal influences simply seem to come out of the blue. Preservation of causality demands instead that, if possible, every force should be attributed to a source from which it originates lawfully.

Leaving the question of intertranslatability aside for a moment, consider Newton's reasoning in favor of a heliocentric account of the planetary system. In describing the phenomena of planetary motion he left it explicitly undecided whether the sun or the earth is located at its center (cf. Newton 1726, 404). This implies that, as regards planetary positions, Newton took a geocentric (Tychonic) account to be empirically equivalent to a heliocentric one. And this is true indeed. Systematically speaking, every periodic motion can be expressed by a Fourier series. In particular, we may use such a Fourier series to represent the motion of the planets around the earth. For that purpose every orbit has to be represented by a superposition of epicyclic motions whose periods of rotation and epicycle radii (i.e., frequencies and amplitudes) are properly adjusted (cf. Kamlah 1971, 208–210).

But, as Newton went on to argue, if we consider the forces that bring about planetary motion and take the possible sources of these forces into account, we are led to the conclusion that the center of gravity of the planetary system roughly coincides with the sun (cf. Newton 1726, 419–421). This means that in light of a dynamical analysis it emerges that only a heliocentric theory can give a consistent account of the causes of the observed motions. In spite of their empirical equivalence, this theory is distinguished with regard to its geocentric competitors by its preservation of causality.

Another case in point is Grünbaum's remetrization of Newtonian time. Grünbaum examines the impact of taking the daily rotation of the earth as time standard. According to the usual interpretation, the time intervals determined this way vary over time because of the tide-produced gradual slowing of the earth's rotation. On the basis of the remetrized "diurnal time" one arrives at an empirically equivalent reformulation of Newtonian theory. One of the peculiarities appearing in this reformulation is, however, that accelerations turn up that are not attributable to any material source (cf. Grünbaum 1973, 66–75). The remetrized variant thus fails to preserve causality. Consequently, in view of our third requirement neither a geocentric model nor the diurnal-time model count as conventional alternatives to the standard

Newtonian theory.

§ 6. Yet there is still too much room for trivializing conventionalism. For there are theories that satisfy these three requirements but are nonetheless irremediably ludicrous. Think of the following two examples: (a) God created the universe (along with all memories, fossils etc.) just five seconds ago. (b) All macroscopic objects evaporate when nobody or nothing registers their existence; and even the most superficial glance makes them recondense instantly. It appears immediately that no empirical facts militate against these somewhat extravagant contentions. Moreover, there is no intertranslatability with respect to the standard interpretations. In the first case, non-theorems of the standard view need not go into non-theorems of the non-standard one. God's omnipotence implies that He could have arranged everything in a quite different fashion. So what is not considered a possible world in the usual physical account may appear perfectly possible in the theologically enriched doctrine. As regards the second case, the evaporation-recondensation mechanism has no analog in the standard view. This case accordingly constitutes an example of deoccamization as explained above. Third, there is no violation of causality involved here. Rather, we have some extra-causes in the two non-standard interpretations. Namely, acts of God, for one (that are invoked here in a larger measure than it's done anyway), and registration as the cause of a body's persistence, for another.

It appears implausible, however, to consider these examples full-fledged cases of conventional alternatives. They are simply too stupid. Cases of this sort are to be excluded by stipulating a fourth requirement, namely, *non-vanishing methodological qualification*. By this I intend that there is at least one criterion from among the class of widely shared methodological values on which the non-standard account fares better. That is, if an account is to pass as a conventional alternative to the received view, it has to qualify as superior in at least one methodological respect. This condition is supposed to capture the intuition that there should be at least a trace of a serious competition between conventional alternatives.

It is at this juncture where the conventionality of distant simultaneity gets discredited in any event. That is, even if intertranslatability among standard and non-standard versions of Special Relativity is denied, the latter are disqualified as truly conventional alternatives on account of their methodological inferiority. What renders the non-standard alternatives unilluminating is the fact that there is nothing to be gained methodologically through the introduction of a deviant synchronization procedure. So it is here that the methodological objections, as recounted in IV.5.§7, become relevant in the end. While it is true that negligible methodological rating leaves the original conventionalist thesis unimpaired, as stressed in IV.5.§7, the thesis itself appears less than impressive.

I call the claim that alternatives satisfying all four requirements actually exist in science *strong conventionalism.* Only by enriching the conventionalist position with a non-synonymy condition and with methodological elements such as those expressed in the requirements (3) and (4), does conventionalism become an interesting point of view.[11] Moreover, I claim that a positive instance of strong conventionalism exists in fact. There is at least one case in which all four requirements are actually met, and this is the case of physical geometry. For the Newtonian as well as for the Einsteinian account of gravitation alternative approaches can be specified that comply with these demands. So let's turn to a presentation of these approaches.

2. NEWTON ROLLED UP AND EINSTEIN STRETCHED OUT: GEOMETRIES AT A DISCOUNT

§ 1. The first item on the agenda is to geometrize ordinary Newtonian gravity. The use of a generally covariant notation is essential to accomplish this task which was first carried out successfully by Elie Cartan in the 1920's. I begin by specifying some properties of Newtonian space-time.

Newtonian space-time is constituted by layers of planes of simultaneity that are ordered by absolute time. Relations of simultaneity and duration are thus independent of the frame of reference chosen. In each simultaneity plane the geometry is Euclidean so that, in sum, Newtonian space-time structure is represented by a family of Euclidean spatial metrics that are temporally ordered by the flow of absolute time. There is no space-time metric g_{ik} (distances are only defined for simultaneous events), but the family of spatial metrics can formally be expressed by a spatiotemporal tensor h^{ij} that bears some resemblance to a truly Riemannian metric tensor (cf. Friedman 1983, 72–77).

Now gravitation is introduced into this space-time. Its characteristics are specified by Poisson's equation that gives a local formulation of Newton's inverse square law and thus represents the field equation of the Newtonian gravitational field. In generally covariant notation Poisson's equation reads:

$$(4) \qquad h^{ij}\frac{\partial^2\Phi}{\partial x^i \partial x^j} = h^{ij}\Phi_{;\,ij} = 4\pi\gamma\rho\,,$$

with Φ representing the Newtonian gravitational potential, ρ the mass density, and γ the gravitational constant. The semicolon denotes the operation of covariant differentiation.[12]

The general covariant formulation of the Newtonian equation of motion applied to motion under the sole influence of gravitation is (cf. Friedman 1983, 94):

$$(5) \qquad \frac{\mathrm{d}^2 x^i}{\mathrm{d}t^2} + \Gamma^i_{jk}\frac{\mathrm{d}x^j}{\mathrm{d}t}\frac{\mathrm{d}x^k}{\mathrm{d}t} = -h^{ir}\Phi_{;\,r}$$

Equation (5) portrays a gravitational force acting in a flat background space-time. This situation conspicuously resembles the scenario presented in VI.1.§1. Comparison of equations (5) and (2) reveals that gravitation is in fact identical to a universal field U^i (as given by Equation (2)). The evaluation of U^i using Equation (5) yields:

$$U^i = -h^{ir}\Phi_{;\,r}.$$

This result invites us to geometrize away Newtonian gravity. Following the recipe outlined in VI.1.§1, gravitational forces are incorporated into the affine connection. Gravitationally influenced trajectories thus appear as geodesics in curved space-time.

$$(6) \qquad \Gamma^i_{jk} = {}^f\,\Gamma^i_{jk} + h^{ir}\Phi_{;\,r}t_j t_k$$

The quantities t_j, t_k denote the inverse of $\mathrm{d}x^j/\mathrm{d}t$, $\mathrm{d}x^k/\mathrm{d}t$, respectively. The new, non-flat connection Γ^i_{jk} is expressed by the flat connection ${}^f\Gamma^i_{jk}$ and the gravitational force per unit mass $h^{ir}\Phi_{;\,r}$. Equation (6) is an implementation of the strategy expressed in Equation (3) (cf. VI.1.§1).

Whereas spatial geometry within each plane of simultaneity remains Euclidean (and spatial curvature correspondingly vanishes), there is now a non-vanishing spatiotemporal curvature. It is given by the geometrized version of Poisson's equation:

$$(7) \qquad R_{jk} = 4\pi\gamma\rho t_j t_k$$

with R_{jk} denoting the so-called Ricci tensor which is obtained from the Riemann curvature tensor. Poisson's equation now expresses space-time curvature as a function of mass density (cf. Misner/Thorne/Wheeler 1973, 289–301; Friedman 1983, 97–102).

Geometrization works in virtue of the principle of equivalence. It follows from this principle that the Newtonian gravitational field is frame-dependent. If one restricts oneself to locally available means, every motion in a gravitational field, as viewed from an accelerated frame, can be interpreted as motion in a different gravitational field, as viewed from an inertial frame (cf. Misner/Thorne/Wheeler 1973, 295; Friedman 1983, 95–97). This means that on the basis of the usual formulation of Newtonian theory, the gravitational field (as well as its potential) and the class of inertial frames (i.e., the flat connection and the corresponding geodesic structure) cannot be determined separately. Rather, the field and the set of distinguished frames reciprocally depend on one another. In the geometrized version, the two underdetermined quantities are replaced by one single object, namely, the non-flat connection (as it appears from Equation (6)).

§ 2. The second instance is provided by the relativistic treatment of the same problem. Whereas Cartan's variant of Newtonian gravitational theory amounts to introducing curvature into flat space-time, the issue is quite the reverse with respect to General Relativity. Einstein's curved space-time is stretched out and flattened. There exist several approaches to this issue; I focus on the version developed by Walter Thirring and crucially improved by S. Deser.

The challenge is to specify an account of gravitation that resembles electrodynamics as closely as possible. That is, gravity is supposed to be represented by a field acting in flat Minkowski space-time. In particular, the potential of the gravitational field is given by a tensorial quantity ϕ_{ik} that is coupled to the stress-energy tensor. Considerations of simplicity and analogy to electrodynamics suggest a Lagrangian which leads to field equations analogous to the linear approximation of Einstein's field equations. The problem is that Thirring's version of the field equations entails unacceptable consequences: The velocity of light comes out

position-dependent – rather than constant – and superluminal velocities are not ruled out. Lorentz invariance is thus destroyed with the result that the whole endeavor of developing a theory of gravitation in analogy to electrodynamics is going to fail. The trouble is fixed by resorting to the following expedient. A coordinate transformation is introduced which locally removes uniform gravitational potentials as well as potentials that depend linearly on the position (i.e., uniform gravitational fields). The result is a changed metric that agrees with the (pseudo-)Riemannian metric of General Relativity.

The physical significance of this maneuver is as follows. Judged on the basis of the original flat metric η_{ik}, a gravitational field tends to deform measuring rods and to distort the running of clocks. In particular, rods shrink under its influence and clocks run slow. The coordinate transformation amounts to compensating and correcting this effect. However, if space-time is measured out with actual – i.e., uncorrected – rods and clocks, then the metric will appear to be Riemannian. Thirring's theory entails that the measured metric g_{ik} is related to the underlying flat metric η_{ik} and the gravitational potential ϕ_{ik} by the equation

$$(8) \qquad g_{ik} = \eta_{ik} - 2f\phi_{ik}$$

where f is proportional to the Newtonian constant of gravitation. This equation is to be understood so that gravity affects the length of rods and the operation of clocks to the effect that the measuring results indicate the presence of g_{ik} – which is in numerical agreement with the Einsteinian metric – rather than the "true" η_{ik}. The same goes analogously with respect to particle motions. Thirring's approach implies that test particles in a gravitational field move along geodesics with regard to g_{ik} rather than η_{ik}. In sum, then, gravity distorts the measuring instruments in such a way that the flat metric becomes a hidden quantity.[13] Thirring's original theory only succeeded in capturing the linear approximation of General Relativity. Worse yet, the theory is inconsistent if the interaction between the material sources of the gravitational field is taken into consideration. It turns out, however, that Thirring's theory can be improved by taking into account that the gravitational field has to be included among the sources of gravitation. This was achieved by Deser who managed to reproduce the full, non-linear Einstein-equations.[14] The result is that Einstein's field equations can be interpreted to the effect that they comply with a universal-force approach. The clue to this interpretation is Thirring's Equation (8) which constitutes an instantiation of the universal-force scheme as specified by Equation (1) (cf. VI.1.§1). This scheme can thus actually be fleshed out and given concrete physical substance.

§ 3. The upshot of the two preceding subsections is that Newton's theory of gravitation can be geometrized and Einstein's can be degeometrized without affecting their empirical consequences. So in both cases empirically equivalent alternatives can actually be specified.[15] Let's now go through the remaining three conditions, as outlined in VI.1.§4–6, so as to see whether we really have full-fledged examples of strongly conventional alternatives.

Cartan's curved space-time model and the traditional Newtonian account are not intertranslatable. This becomes clear by focusing attention on Equation (6) which constitutes the crucial obstacle to intertranslation. This equation can be interpreted as the definition of the Cartan-connection Γ^i_{jk} in terms of the flat Newtonian connection ${}^f\Gamma^i_{jk}$ and the gravitational force. The transition from Newton to Cartan is thus unproblematic. Difficulties emerge, however, with the reverse transition. For there is no way to define either the force or the flat connection by relying exclusively on Cartan-concepts. Equation (6) makes it clear that the definition of either one of the Newtonian quantities must resort to the complementary one. Accordingly, Cartan's theory is not interpretable in Newton's, and this is why intertranslatability fails.

The same goes for the relation between the Einsteinian account and the Thirring–Deser alternative. A look at Equation (8) reveals that the conceptual situation is exactly analogous to the case of Newton and Cartan. While Equation (8) provides a definition of the Einsteinian g_{ik} in terms of Thirring's η_{ik} and ϕ_{ik}, the reverse transition is blocked. Neither the fixed background metric η_{ik} nor Thirring's gravitational potential ϕ_{ik} is definable in General Relativistic terms. Apart from special instances, General Relativity does not allow for splitting up the metric into a flat background and a variable part. As a result, it is not interpretable in the Thirring–Deser account so that intertranslatability does not obtain.

As regards causality, the field equations of the two alternative approaches preserve the causal relations between the field quantities and their sources. In the Cartan variant, Equation (7) is the precise analog of Poisson's equation in its usual form (4). With respect to causality it makes no difference whether masses are supposed to create a universal field or a variable curvature. The same holds for the Thirring–Deser model. Thirring's field equations relate the ϕ_{ik}-field to its sources.[16] Moreover, if the self-interaction of the gravitational field is actually taken into consideration, one arrives at the full Einstein equations and, consequently, at the causal relations inherent in them. In fact, this feature constitutes the crucial advantage of the Thirring–Deser approach over the Reichenbach account. The former, in contrast to the latter, allows for treating gravitation as a universal force in a way that is causally as kosher as geometrized gravity is. The Thirring–Deser approach is not restricted to kinematics; rather, it succeeds in accommodating the possible causes of the gravitational field as well. The approach incorporates a dynamical account of the gravitational field and preserves causality to the same extent as standard General Relativity does.

§ 4. Let's now address the crucial aspect, namely, methodological qualification. It has been argued by Glymour that the two universal-force accounts are, from a methodological point of view, vastly inferior to their geometrized counterparts. Glymour focuses on the testability of the models involved, and his argument runs roughly as follows. Every empirical test of the two geometrized accounts also constitutes a test of the corresponding degeometrized analogs – but not the other way around. For the latter contain some untested elements that are absent in the geometrized versions, namely, equations (6) and (8) respectively. The degeometrized quantities figuring in these relations cannot uniquely be determined on the basis of the pertinent theories. Consequently, the relations themselves cannot be tested. This means that the geometrized variants are better tested than their universal-force analogs and are thus to be preferred for methodological reasons (cf. Glymour 1977, 238–245; Glymour 1980, 358–366).

Glymour's argument doubtlessly has some force and, in addition, its conclusion is non-committal as to the particular methodological criterion of testability that Glymour invokes. For geometrizing gravitation amounts to replacing two frame-dependent quantities by a single frame-independent quantity, and such a move is certainly suited to score a methodological point – whatever the particulars of one's pet methodology are like.[17] So if the four accounts at issue are considered in isolation, the respective geometrized versions certainly qualify as methodologically superior to their universal-force counterparts.

But this is not the end of the story. After all, one may ask why lots of (to all appearances) serious people tried to develop a degeometrized version of General Relativity if that's really such a silly enterprise. In fact, there is an additional methodological property operative here that in turn diminishes the merit of geometrization. This property concerns the integration of such an account into the remaining parts of scientific knowledge. So we change now the methodological perspective and examine the coherence, or lack of coherence, that is generated in the overall picture of physical interactions by geometrizing or degeometrizing gravitation.

Let's suppose Cartan (or somebody else) had come up with geometrized gravity half a century earlier, that is, in a time with Newtonian gravity ruling unchallenged, Riemannian geometry invented, and Maxwell's equations formulated. Is it plausible to presume, the question is, that such a scenario would have ended up with the eventual demise of Newtonian gravitation and its substitution with Cartanian geometrization? I strongly doubt that. I rather guess that one of the objections raised against the latter would have been that the beautiful internal virtues of a geometrized account of gravitation – as outlined above – are purchased at the expense of an ugly external vice. For adopting Cartan gravitation would have created a serious incoherence with respect to the only other physical interaction known at the time, namely, electrodynamics. Underlying electrodynamics is the view of electromagnetic forces acting in a fixed background space-time. And since electromagnetic forces per unit mass depend on particle properties (namely, charge and velocity), there is no electromagnetic analog to the principle of equivalence. Hence electrodynamics cannot be geometrized.[18] Accepting Cartan's theory would thus have induced a considerable decrease in the conceptual unity of scientific knowledge.

The upshot of this discussion can be put a bit more systematically by generalizing its moral in light of some methodological considerations of Larry Laudan's and Thomas Kuhn's. Laudan emphasizes that the methodological appraisal of a theory should not be based – and as a matter of historical fact has not been based – exclusively on that theory's empirical record. There is an additional, conceptual level of theory evaluation. If a theory generates conceptual problems, its overall appraisal measure decreases. One example of such a conceptual problem is external incoherence. This means, if there is a tension or conflict between the basic tenets of a theory and some other well-founded account, the methodological rating of the theory is reduced (cf. Laudan 1977, 50–54, 68).

One recognizes immediately that our fictitious Cartan case instantiates this rule. Whereas Cartan's approach to gravitation and Maxwell's approach to electrodynamics do well when considered in isolation, a tension arises between them when viewed in combination. For it is simply implausible that the only two interactions known at the time should be accommodated by conceptually completely disparate accounts. And since conceptual unity can be reestablished easily by degeometrizing gravitation, the methodological balance is redressed in favor of the traditional Newtonian view.

The same point can be made on the basis of a Kuhnian argument. Unlike Laudan, Kuhn is not concerned with incorporating conceptual problems into a general rule for theory appraisal. Rather, he denies that any such rule can sensibly be specified. This claim is supposed to be established by giving, among other things, a collection of generally shared standards for evaluating theories – such as accuracy, internal consistency, external coherence, broad scope, simplicity and the like. These standards, so the argument runs, do not allow for unambiguous theory choice decisions because they may be in conflict with one another when being applied to particular cases. For example, one theory may be simpler whereas its competitor may stand out due to its wider domain of application (cf. Kuhn 1977, 321–322).

This general analysis fits well with our fictitious case study. Whereas Cartan's account excels regarding internal qualities such as determinateness of the theoretical entities it invokes or regarding the testability of the laws it specifies, it fares rather poorly, by contrast, with respect to external coherence. And this deficiency is detrimental to the methodological ranking of Cartan's account. So we may draw the following conclusions from this discussion. First, external coherence belongs to the class of generally accepted methodological criteria and, second, in the case at hand it favors a universal-force theory of gravitation. Consequently, this case constitutes an instance of strong conventionality.

§ 5. The very same reasoning applies to the relation between Einstein's approach and the Thirring–Deser variant. So we no longer treat a fictitious scenario but now turn to a real-life example. The internal methodological merits of standard General Relativity have already been outlined; what remains to do is to point out the external incoherence generated by that approach. This incoherence emerges with respect to Quantum Field Theory. All physical interactions other than gravitation are described in terms of quantum fields, and the design of the corresponding theories is completely at variance with General Relativity. In quantum field theories every interaction is regarded as the effect of an exchange of bosons between the particles involved. All interactions are thus produced by quantized fields, and the whole process takes place in flat Minkowski space-time. No trace of geometrization appears here.

Moreover, the external incoherence created here is even more marked than the conceptual tension emerging in the Cartan example. For General Relativity and quantum field theories are outright incompatible and not merely jointly implausible. The reason is that Einstein's field equations contain the stress-energy tensor which expresses the material sources of gravitation in a classical fashion. But no classical description of matter can claim validity anymore; the structure of matter certainly belongs to the realm of quantum mechanics (cf. Wald 1984, 382). This entails that Quantum Field Theory and General Relativity cannot both be true. There is no way to a peaceful coexistence; a shoot-out is inevitable in the end.

It is against the background of this theater that the Thirring–Deser account becomes important. On the basis of this account, gravitation appears as the classical field that corresponds to quantum mechanical particles with zero rest mass and spin two (gravitons). Einstein's field equations are reinterpreted as a theory of a self-interacting spin-two field in Minkowski space-time. Accordingly, General Relativity becomes the classical limit of a quantum field theory of gravitation (cf. Misner/Thorne/Wheeler 1973, 435; Deser/van Nieuwenhuizen/Boulware 1975, 7; Wald 1984, 383). It leaps to the eye that such an account of gravitation is much more intimately integrated into the remainder of contemporary physics. As regards external coherence, the Thirring–Deser theory is superior to standard General Relativity.

It is striking, moreover, that precisely this theme of the unity of science is the sought-for driving force behind the attempts to degeometrize General Relativity. Thirring, for instance, points to the discrepancy between the quantum field picture of interactions and the entirely different Einsteinian approach. He is explicitly committed to the objective of unifying the field theories by removing the "mysterious special status of General Relativity" (cf. Thirring 1959, 79, 83). Like-minded arguments are advanced by Steven Weinberg in order to justify his purely instrumentalist interpretation of General Relativity. Geometrizing gravitation, he contends, has driven a wedge between gravity and particle physics. The geometric concepts are thus to be viewed as merely mathematical means for squeezing some physical stuff out of the principle of equivalence (cf. Weinberg 1972, VII–VIII). Space-time curvature thus assumes the status of a mere *façon de parler*.

It is to be admitted, on the other hand, that the Thirring–Deser account does not itself constitute a quantum theory of gravitation. Worse yet, it cannot even be transformed into such a theory. The problem is that the attempts to quantize the gravitational field on the basis of this account lead to a non-renormalizable theory. Accordingly, the Thirring–Deser account as it stands is unsuited as a viable starting-point for developing a full quantum theory of gravitation.

Disastrous as it is with respect to physics, this result does not interfere with the philosophical moral I wish to extract. On the contrary. I am confident that a really working quantum theory of gravitation will predict some new results that differ from the empirical consequences of General Relativity. In that case we would no longer deal with empirically equivalent alternatives. I trust that adopting the future quantum theory of gravitation won't be a matter of conventional choice.

Returning to the situation as it actually is we realize that in spite of the internal shortcomings of the Thirring–Deser account, there is one important reason that speaks in its favor. Stripped to the essentials, standard General Relativity is inconsistent with Quantum Theory, and the odds are far better for the latter than for classical physics to pass as the comprehensive account of nature. In the usual picture gravitation appears as the last remnant of classical physics reaching anachronistically into the quantum age. St. Patrick's cathedral amidst Fifth Avenue business towers. It's a good bet that the eventual quantum theory of gravitation will bid farewell to geometrization, and this is what supports a universal force account of gravity.

This situation can be analyzed in terms of the concepts developed in the preceding chapters. As indicated repeatedly, the universal-force scenario exhibits the structure of a Reichenbach loop. Furthermore, the intractability of this geometric Reichenbach loop between the evaluation of the universal force and the prevailing geometry derives from its theoretical isolation (cf. IV.4.§4). None of the quantities looming in the loop can uniquely be specified using the corresponding network of laws. This judgment, however, is called into question by the present analysis. For the present argument comes down to employing Quantum Field Theory for the external fixing of the structure of space-time; it has to be Minkowskian in kind. As a result, gravitation becomes a universal force. If the isolation of the geometric Reichenbach loop is removed by integrating it into the wider nomological framework of quantum theory, it can be resolved. A bit of social life sometimes helps.

The conclusion is that we are indeed presented with a full-fledged real-life case of strong conventionality. The Thirring–Deser account receives a non-vanishing methodological rating since there is indeed one criterion among the class of generally esteemed values in light of which this account fares better than the standard approach. Accepting or rejecting geometrized gravitation is thus a conventional choice in the strong sense. This does not imply that such a choice must be made arbitrarily and without good reasons. After all, I just discussed the relevant pros and cons, and a weighting of arguments is, of course, not a blind choice. But conventionalism never claimed that concerning conventional alternatives we are in a Buridan's ass situation. It is perfectly possible to make conventional choices in a justified manner. The point merely is that they should be recognized as choices whose outcome is dictated neither by nature nor by elementary constraints of reason.

§ 6. Accordingly, the claim is that conventionalism can be detrivialized but still be instantiated by going half-way in the direction of the realist counterarguments. That is, I admit, first, that if theoretical alternatives advanced by conventionalists fail to comply with minimum standards of reasonableness and plausibility, they should not be taken seriously. A conventionalism of that sort is self-defeating. But I claim, second, that conventionalism can be given a higher profile by incorporating such standards into it. The position emerging this way turns out to be supported by the case of physical geometry.

So, after all, the early conventionalists were on the right track in singling out physical geometry as the paradigm example of their contentions. Although they overshot their goal by claiming that conventional alternatives are entirely on a par epistemically, they indeed smelled out the right case. If judged in terms of the criteria proposed here, physical geometry really constitutes a case for strong conventionalism. Physical geometry is conventional in a particular sense; it is more conventional than other physical theories. In addition, the conventional aspects involved really matter. For as stressed in IV.4.§4, it certainly makes an interesting difference whether gravitation is supposed to be a physical force analogous to the remaining interactions known in physics, or whether it is granted the special status of representing the structure of space-time. If science is supposed to direct us on what there is, the attitude one takes on the geometrization issue has a bearing on what one assumes the physical world ultimately to be made of.

NOTES

NOTES TO CHAPTER I

1. This sketch of the double-language model is based on Hempel 1952, 20–39; Carnap 1956, 38–52; Hempel 1958, 177–179, 182–185; Nagel 1961, 79–105.
2. Cf. Kuhn 1962, 110–127. For an epistemological clarification and elaboration of this Kuhnian claim see Hoyningen-Huene 1989, 41–71.
3. Cf. the letters exchanged between Newton and Flamsteed during the fall 1694 and especially Newton's letter to Flamsteed dated Dec. 20, 1694 (in Scott 1967, 61–62). During that period Newton was actively working at the second, improved edition of his *Principia* and he was in urgent need of accurate (i.e., corrected) values of the planetary positions as support of his gravitational theory.
4. The following outline of affinity theory is based on Carrier 1986a. The theory was already mentioned in the introduction.
5. This procedure was interpreted as a quantitative refinement of the method of substitution reactions as referred to in the introduction. Substitution reactions only furnished a rank-ordering of affinity strengths.
6. For Dalton's atomic theory cf. Dalton 1808, 211–216, 218–220; for a more detailed description of it see also Nash 1967, 14–31. I will go into the underdetermination problem in II.4.§3.
7. This feature has been emphasized especially by Lakatos. Lakatos drew the conclusion that a conflict between theory and evidence is to be construed as a conflict between two theories, namely, explanatory theory and observation theory; cf. Lakatos 1970, 44–45.

NOTES TO CHAPTER II

1. Cf. Carrier 1989 for the introduction of this concept.
2. This is called a "bootstrap-procedure" for testing hypotheses; cf. II.4.§1.
3. For a more thorough discussion of this feature see Carrier 1991a, 48, 50; see also V.4.§1.
4. This introductory sketch is largely based on the presentation in Balzer/Moulines/Sneed 1987; cf. esp. 26–47; 168–170.
5. This feature merely suggests and does not establish the presence of potential strong completeness because the latter requires in addition that necessary corrections be feasible on the basis of the hypothesis in question. But corrections are left out of consideration at present.

6. This is somewhat inaccurate historically since Sneed never gave a concrete analysis of precisely this situation. Even so, it seems to me that it is situations of this sort that express the philosophically interesting aspects of Sneed-theoreticity.
7. Structuralists believe, by contrast, that every test of a hypothesis presupposes the concrete evaluation of all quantities occurring in this hypothesis; cf. Sneed 1979, 37; Stegmüller 1979, 17–18. It is clear, incidentally, that the method described merely yields ratios of the values of theoretical quantities. Yet the remaining indeterminacy only concerns units, and certainly nobody will demand that the choice of units be dictated by experience.
8. Bold types express vectorial quantities.
9. The application of the extensivity constraint presupposes that the equality of the mass of different bodies can be ascertained. This evidently requires a rudimentary sort of instantiation. The procedure described is hence not entirely non-instantiative in character.
10. Cf. Glymour 1980, 31. The precise impact of this problem is still under debate; cf. Grimes 1990, 521.
11. This condition is inspired by Grandy's concept of "selective confirmation" which he invokes to dispel Hempel's paradox of confirmation; cf. Grandy 1967, 23; cf. also the modified version in Glymour 1980, 157–158. This discriminative nature of trustworthy tests is likewise stressed by Mayo 1991, 531–532, 543–547.
12. It seems even doubtful whether the central theory-element is really always used as a criterion of adequacy for the application of special laws. After all, a special law may be the specialization of several distinct theory-elements. Snell's law of refraction can be obtained likewise from a corpuscular theory and a wave theory of light. Special laws are more independent of theoretical cores than structuralism would have it.
13. "Non-mechanical" means here that mechanics is not used as observation theory, notwithstanding its possible invocation as criterion of adequacy. Basing force-measurements on Lorentz's force law, i.e., on an electrodynamical law, is a non-mechanical procedure in this sense.
14. Cf. Glymour 1980, 110, 114–117, 126, 174; cf. also Lambert/Brittan 1987, 99–100. As regards the ancestry of the bootstrap model, the central idea of the first condition, namely, that the confirmation of a hypothesis relies on its instantiation, is due to Hempel; cf. Hempel 1945, 36–38. The second demand that any experiment that is to pass as an empirical test must permit the occurrence of counterevidence has a markedly Popperian ring.
15. The term "deoccamization" refers to Ockham's razor according to which entities ought not be multiplied without necessity. The complex expression of predicates neither of which has any theoretical or empirical import falls victim to this instrument.
16. Cf. Glymour 1980, 112, 122–123; the concordance requirement is taken from Weyl 1949, 121–122. The bootstrap model represents a formal account of confirmation and construes scientific theories as axiomatic structures formulated in first-order

language. It is designed for assessing the bearing of evidence on hypotheses in logical terms. For this reason there is a lot of technical fine print contained in the model which I have omitted here.

17. I reconstruct this reasoning from a passage of Thomson's quoted in Glymour 1980, 243; the general procedure applied is outlined *ibid.* 230–232. The background of Dalton's views is provided by Nash 1967, 24–27. Moreover, I simply take modern combining weight ratios in order for not to get lost in the diverging results of various researchers and in the different scale conventions.
18. For Cannizzaro's method see Cannizzaro 1858, 3–21; Nash 1967, 109–112.
19. This condition was considered and rejected by Hempel who objected that it led into the irrelevant conjunction problem; cf. Hempel 1945, 31–33. This flaw can be remedied, however, by demanding essential use of the hypothesis in the derivation under consideration; cf II.3.§5.
20. It has also been observed by Edidin that Glymour's explicit computation of the constant can be dispensed with; cf. Edidin 1983, 51. But the use Edidin makes of this observation is completely different from mine. He believes that every use of a hypothesis for calculating a quantity occurring in it implies the guaranteed fit of the resulting values with experience; and he concludes from this that using a hypothesis for this purpose should be prohibited; cf. *ibid.* 50–51. Glymour endorses this conclusion; cf. Glymour 1983, 5. But the problem Edidin sets out to solve does not in fact exist, and accepting his proposal would deprive the bootstrap model of what seems to me to be one of its most interesting facets.
21. Kirchhoff's first law states that the sum of all currents at a junction in a network is zero. This implies that the current intensity remains unchanged along either branch. The second law says that the sum of all potential drops along a closed path in a network is zero. It follows that the potential drops across the two branches are equal.
22. There exist variants of this procedure which equally serve the desired purpose. We may, for instance, ascertain the initial water temperature using a second thermometer – assuming the two to be equally reliable.
23. At least let's suppose for the sake of argument that they are. Let's suppose, that is, that repeated recourse to the same quantity is in fact necessary, or, in other words, that no further, independent correction methods exist. If there is such a cheap way out of the correction problem the whole scenario loses its point. And since we are concerned with clarifying the conceptual structure of correction problems we should consider ourselves restricted to the methods outlined.

NOTES TO CHAPTER III

1. Bold types refer to vectorial quantities.

2. Cf. Bohr 1936, 292; Bohr 1961, 311–313. The first step of the argument expounds the so-called "disturbance theory" of measurement while the second step amounts to arguing that in light of the verificationist semantics, as outlined in I.1.§1, the disturbance theory is self-defeating. In the mean time, the disturbance theory has turned out to be untenable for purely physical reasons; cf. Herbert 1985, 160, 228; Redhead 1987, 67–68.
3. In fact, von Neumann represented a quantum mechanical state by a vector rather than a function but this difference is immaterial in the present context.
4. Here and in the following a normalization factor is omitted for the sake of simplicity.
5. There are no states combining the occurrence of a "+"-value in the object with the occurrence of a "−"-value in the instrument. This expresses the demand that we are dealing with an ideal instrument: A given object state is always tied to the corresponding instrument state.
6. For this presentation of the many-worlds interpretation cf. DeWitt 1970, 33–34; Jammer 1974, 508–517; Rae 1986, 75–83.
7. For this criticism cf. Ballentine 1973, 232–233; Bell 1976, 14–15; Albert/Loewer 1988, 201–203.
8. Albert and Loewer try to fix the shortcomings of the many-worlds interpretation by resorting to even more bizarre conceptual means. They advocate, namely, the "many-minds view" according to which every person is endowed with an infinity of distinct minds each of which corresponding to a possible measuring result; cf. Albert/Loewer 1988, 203–210.
9. An argument to this effect was advanced by Kant in support of his thesis that psychology can never become an experimental science; cf. Kant 1786, 8.
10. Cronbach/Meehl 1956, 177–178. Cronbach and Meehl played a leading role in developing the concept of construct validity. See Lienert 1969, 261–264 and Suen/Ary 1989, 177–179 for additional descriptions of construct validity.
11. On the other hand, such Duhemian test problems occur with a vengeance here. If a given test does not show the presumed correlations there is no general recipe for deciding whether the test's construct validity is wanting or whether the underlying theory is false. This Duhemian problem lies at the bottom of Suen/Ary's complaint about the "frustrating circular nature of [the] logic" of construct validity; Suen/Ary 1989, 179.
12. Cf. Freud 1900, 322–323; later qualified as a result of empirical pressure; cf. *ibid.* 270–271.
13. This criterion is applied by Freud in a striking fashion in analyzing a "Freudian slip"; cf. Freud 1901, 11. For another application cf. Meehl 1983, 403.
14. The unity of scientific method is stressed in Carrier/Mittelstrass 1990, 25–28.
15. Cf. Freud 1900, 528. This argument strongly resembles the so-called "Miracle Argument" advanced in support of scientific realism; for a reconstruction cf. Carrier 1991b, 23–24, 28.
16. Cf. Grünbaum 1983, 327; Grünbaum 1984, 235, 277. Freud recognized this difficulty and tried to limit its impact: In a successful therapy only the correct

suggestions are eventually accepted by the patient. The emerging diagnosis is thus correct despite the therapist's admitted suggestive influence. For different reconstructions of Freud's line of argumentation cf. Grünbaum 1984, 130–147; Richardson 1990, 672–676; Derksen 1992.

NOTES TO CHAPTER IV

1. For a sketch of some of the properties of non-Euclidean geometries see Reichenbach 1928, 1–8; Sklar 1974, 16–25.
2. The only exception concerns those effects of a possibly present external embedding that do not express themselves in the internal geometry. Consider a two-dimensional example. A plane and a cylindrical surface have the same internal geometry and only differ with respect to their corresponding embedding in a three-dimensional space. By contrast, the difference between a plane and a spherical surface manifests itself in a different internal geometry, i.e., a different metric tensor.
3. In this way the concept of differentiation is introduced into a Riemannian manifold. Parallel displacement means that the direction is maintained, and this can be put thus that the derivative vanishes. A generalization of this approach provides a definition of the so-called covariant differentiation of a vector.
4. As it was first pointed out by Weyl, this possibility uniquely characterizes Riemannian space. Only in Riemannian spaces does the metric unambiguously determine one affine connection; cf. Weyl 1923b, 46–47.
5. It is noteworthy that thermal and gravitational influences are treated here on a par, namely, as distortions that are to be corrected. As it will be seen, the situation is quite different in General Relativity; cf. IV.3.§5.
6. For this argument cf. Poincaré 1902, 88–91; cf. also Grünbaum 1973, 18–21; Kanitscheider 1971, 180–181.
7. For an outline of the basics of affine and metric geometry see Weyl 1923a, 14–20, 24–26; Friedman 1983, 32–43.
8. Cf. Einstein 1918, 241–242; Einstein 1922, 55–56. As it turned out later, Mach's principle is actually not satisfied by full-grown General Relativity. The principle thus provided a heuristic impulse to the theory's development, but it had to be discarded upon completion of the theory.
9. A more fundamental remedy is available within the framework of developed General Relativity; see § 5.
10. Einstein's line of reasoning is reconstructed in Norton 1985, 222–223.
11. Cf. Kretschmann 1917, 576–579; Friedman 1983, 55; cf. also VI.2.§1. However, the principle of general covariance that Einstein actually invoked is different from the modern version (and the one employed by Kretschmann). As Einstein used the principle, it is not devoid of physical content; cf. Norton 1989, 1241–1242, 1255.

12. Cf. Weinberg 1972, 68–69; Misner/Thorne/Wheeler 1973, 386; Stephani 1980, 66. For qualification of the local version of the principle see § 7.
13. Cf. Einstein 1917, 65–66; Einstein 1922, 59–61. The geometry on a rotating disk is extensively discussed in Grünbaum/Janis 1979.
14. For Einstein's identification of the affine connection with the gravitational field see Einstein 1916, 143. This identity has to be carefully formulated though. It is certainly inappropriate to interpret those changes in the affine connection that arise merely from a change in the coordinate system chosen, and without a concomitant change in the observer's state of motion, as changes in the gravitational field strength. This can be expressed by making use of a coordinate-free representation of geometric quantities. I won't go into these technicalities here.
15. It deserves notice that the geodesic structure is frame-independent (as it must be). Namely, although the vanishing of Γ^i_{jk} is frame-dependent, Equation (4) is not; and it is this equation that fixes the geodesic or affine structure. So, whereas the affine connection is frame-dependent, the affine structure is frame-independent.
16. Strictly speaking, it is not the full Riemann curvature tensor that figures in the field equations but its contracted form, the so-called Ricci-tensor generated from the former by taking averages.
17. For an account of tidal forces in the framework of General Relativity cf. Misner/Thorne/Wheeler 1973, 860–862.
18. This objection has been put forward as early as 1911 by Laue in a letter to Einstein; cf. Norton 1985, 214.
19. A causal interpretation can in general only be upheld if it takes two factors into account; First, that the gravitational field itself is among the curvature sources; the vanishing of the stress-energy tensor alone does not imply the vanishing of the Riemann curvature tensor. Second, the influence of the curvature on the distribution of matter (as instantiated by geodesic deviation) and thus on its material sources has to be factored in. So, the relation between curvature and its sources exhibits some intricate aspects; it cannot adequately be portrayed as straightforward unidirectional dependence.
20. It will be seen that in reality this is not that easy; cf. V.2.§1.
21. This claim will be fleshed out in VI.2.§2.
22. Note that goal and gist of this rule agree with Carnap's criterion as outlined in IV.2.§3.
23. Cf. Reichenbach 1928, 34–35. Note the similarity to the views of Poincaré (cf. IV.2.§2).
24. In general, the situation is governed by Buckingham's theorem according to which all basic physical laws can be reduced to relations among dimensionless quantities. This theorem implies that a numerical change in a quantity with a dimension has to be compensated by a suitable adjustment of the values of other quantities related to the one in question in order to keep the dimensionless products of Buckingham's theorem constant; cf. Dieks 1987, 384–385.
25. And the same goes analogously for some alternative rule.

26. The distinction between relative confirmation and real confirmation has been developed and stressed in Edidin 1988.
27. In VI.2.§5 I will discuss a strategy possibly fit for removing the theoretical isolation of the geometric Reichenbach loop. Relying on this strategy the loop can in principle be dissolved through external fixing.
28. Cf. Reichenbach 1928, 263. Tidal forces are universal three-forces that are not set equal to zero by standard General Relativity. That is, their influence is considered as a distortion and is corrected (cf. IV.3.§7). Tidal forces are only set equal to zero in the loose and somewhat metaphorical sense of attributing their origin to curvature.
29. Cf. Reichenbach 1928, 114–119. This claim has been confirmed by Grünbaum. Grünbaum explored the possibility of introducing an alternative standard of equality for temporal intervals into Newtonian mechanics; a standard, that is, non-linearly related to the usual one. Although, admittedly, the alternative standard leads to a more cumbersome description of the facts, it is empirically equivalent with the received one; cf. Grünbaum 1973, 68–77.
30. Cf. Reichenbach 1928, 123–128. Note that Reichenbach's conventionality of distant simultaneity has nothing to do with the relativity of simultaneity. The latter occurs between different inertial frames, whereas the former arises within one inertial frame. After all, all gadgets referred to in the experiment are at relative rest.
31. In the most general case ε may vary with the position as well.
32. Cf. Reichenbach 1928, 129–135; Reichenbach 1938, 127–129. I will expand on the adequacy of this assessment in the following section.
33. Cf. Salmon 1977. For a more recent reconfirmation of Salmon's analysis see Clifton 1989.
34. This interpretation is to be taken with a grain of salt since according to Special Relativity all inertially moved clocks are on the same footing epistemically. In particular, each of the two clocks in relative inertial motion lags behind its counterpart. Relativistic time dilation is symmetric, and there is, consequently, no privileged state of motion which could represent the true value. In the present argument, however, this aspect can be set aside. We consider the situation with respect to the rest frame of C_A and C_B.
35. Cf. Bridgman 1962, 65–67; Ellis/Bowman 1967, 130. Ellis/Bowman specify an additional procedure for establishing a unique transport synchronization; cf. *ibid.* 129–130. See also Salmon 1977, 265–266; Carrier 1989, 424–425.
36. Cf. Salmon 1969, 46–49; the proof is given on the basis of the exact time dilation formula, not by using its approximate analog, i.e., Equation (10).
37. A general expression for one-way ε-velocities is derived in Winnie 1970, 85. On this expression, equal standard velocities come out as different ε-velocities.
38. Cf. Reichenbach 1928, 127; Grünbaum 1973, 77; see also IV.4.§1.
39. This point has been forcefully made by Havas; cf. Havas 1987, 444–445.

NOTES TO CHAPTER V

1. The considerations of the present chapter are to a certain extent improvements on Carrier 1990; Carrier 1993.
2. Actually, it follows from the weak principle of equivalence that free fall motion is unique in the just-given sense. But the Ehlers-Pirani-Schild program of course forbids invoking this principle at the present stage of the construction.
3. For this procedure cf. Ehlers/Pirani/Schild 1972; Ehlers 1973b, 23–32; Ehlers/Köhler 1977, 2014, 2017; Ehlers 1988, 148–158; Heilig/Pfister 1990. In fact, this procedure does not end up with precisely the space-time structure of General Relativity but with a slightly more extended class of metrics, namely, a so-called Weyl geometry. But this problem need not concern us here.
4. Measuring the relative change is the correction procedure suggested by Marzke-Wheeler; cf. Marzke/Wheeler 1964, 49, 53. This correction procedure only fails if the clock is to be incorporated into the Ehlers-Pirani-Schild program of founding space-time measurements on simple experiences.
5. In fact, this feature underlies the Marzke-Wheeler proposal as sketched in § 4.
6. Cf. Ohanian 1977, 906. This ratio thus belongs to the class of effects, already addressed in IV.3.§7, that furnish a local measure of curvature.
7. This ratio is given by the product of the tidal force and the squared period of operation; cf. Ohanian 1977, 906.
8. This limitation can be gotten rid of by recourse to a differently constructed (but like-inspired) clock, namely, the *Perlick clock.* The basic idea is that the acceleration of freely falling particles assumes distinguished properties if it is evaluated using a standard clock. In particular, if such a clock is itself freely falling, then the trajectories of nearby freely falling particles, as recorded by the radar coordinates (see § 2), should come out unaccelerated. That is, provided that the emission and return times of the light pulses employed for keeping track of free-particle trajectories are measured by a standard clock, these trajectories become geodesic. This criterion can then be applied for singling out the right type of clock. The essential point is that it is possible to generalize this procedure to non-geodesic clock motion. Roughly speaking, the deviation of free-fall particle motion from the corresponding geodesics is taken as indicating the clock's own non-geodesic motion. This evaluation is subsequently used for a rectification of the clock's running. On the Perlick procedure, the adequacy of arbitrarily moved clocks can thus be assessed; cf. Perlick 1987, 1068–1069.
9. Marzke/Wheeler 1964, 62. If conjoined to like-minded passages in Marzke/Wheeler 1964, 48, this quote indicates these authors' commitment to a notion akin to Einstein-Feigl completeness. A like-inspired remark is also found in Misner/Thorne/Wheeler 1973, 393.
10. Einstein 1949, 677. A preliminary stage of the argument is found in Einstein 1921, 122–123.

11. Cf. Reichenbach 1951, 135–137. There Reichenbach merely restates the position under attack.
12. Cf. IV.3.§6. For the stress-energy tensor see, e.g., Weinberg 1972, 43–46; Landau/Lifshitz 1976, 93–103.
13. Grünbaum is well aware of this limitation; cf. Grünbaum 1973, 146.
14. On the concept of axiomatization see, e.g., Sklar 1977, 207.
15. It was this failure of space-time theory to accommodate the internal operation of usual rods and clocks which motivated Einstein's call for more complete theories; cf. II.1.§2.

NOTES TO CHAPTER VI

1. Equation (1) is identical to Equation (5) in IV.4.§3.
2. Equation (3) is equivalent to Equation (8) in IV.4.§5.
3. The empirical criteria for inertial motion – as specified in V.1.§2 – do not detract from this possibility. These criteria only serve to implement into experience the *decision* to regard test-particle motions in a gravitational field as inertial.
4. This connection between conventionality and non-realism is emphasized in Ben-Menahem 1990, 261.
5. See also Carnap's remetrization procedure as discussed in IV.2.§3.
6. Duhem 1906, 171. Cf. also Ben-Menahem 1990, 265–266.
7. The need for a non-triviality criterion of that sort has been stressed in Ben-Menahem 1990, 269.
8. Cf. Bonevac 1982, 65–70. I have given a simplified account of the interpretability-concept; for the present purpose it suffices this way. Note that interpretability does not require every T_1-theorem to be mapped onto precisely one T_2-theorem.
9. This notion of mutual interpretability approximately coincides with the synonymy conditions advocated in Glymour 1977, 229–230 and Putnam 1983, 38–39. Glymour rightly stresses that the translation rules have to establish generally covariant relations. Putnam's condition of "relative mutual interpretability" is weaker than mutual interpretability *simpliciter*. It considers mutual interpretability in a restricted range of initial and boundary conditions sufficient for synonymy (on the concept of relative interpretability see, e.g., Carrier/Mittelstrass 1991, 44–49). On this attenuated version, however, theories which are not empirically equivalent may come out as synonymous.
10. Cf. Winnie 1970, 85. This equation is an immediate consequence of Equation (9) in IV.4.§6.
11. The introduction of methodological criteria into the procedure of judging what is to count as conventional alternatives was foreshadowed by Reichenbach. In his well-known "cubical-world" example, Reichenbach discussed two empirically equivalent hypotheses that in his view should not pass as conventional alternatives.

His argument is that one of the hypotheses invokes a causal mechanism known from other instances and the other one does not. This feature distinguishes the first one as being more likely to be true; cf. Reichenbach 1938, 122–124. This means that Reichenbach tacitly employs the methodological criterion of uniform explanation and assumes that this criterion is suited to give epistemic primacy to one of the rivaling accounts – in spite of their empirical equivalence. Accordingly, the brand of conventionalism imputed to Reichenbach in § 1 was qualified by him in his later work. He maintained the view, though, that the conventionality of distant simultaneity still constitutes an instance of this qualified version; cf. *ibid.* 128–129.

12. Covariant differentiation is defined so that it preserves tensorial nature; that is, the covariant derivative of a tensor is again a tensor. In the local inertial frame covariant differentiation reduces to ordinary partial differentiation.
13. For this sketch of Thirring's theory see Thirring 1959, 88–98; Thirring 1961, 97–103; Sexl 1967, 274–278; Sexl/Urbantke 1983, 320–330.
14. Cf. Deser 1970, 9–13; cf. also Wyss 1965; Sexl/Urbantke 1983, 330.
15. A possible objection to the claimed empirical equivalence runs as follows. In a universe with a positive overall curvature it should be possible to send a light signal or a spaceship around the whole universe – just as it is possible to travel around the earth. Such a round-trip, however, is ruled out on the flat account. But as a matter of fact, our universe happens not to be of this kind anyway.
16. Glymour objects that these field equations are not instantiated (cf. Glymour 1977, 243), i.e., that there are no objects satisfying these equations. But in fact there are. Since Thirring's field equations largely agree with the linearized Einstein theory, the former are instantiated to the same degree as the latter. The vacuum propagation of gravitational waves, for example, constitute such an instantiation.
17. I already pointed to the gain in conceptual parsimony achieved by geometrizing gravitation in IV.3.§8.
18. It is true that one can perform such a geometrization by introducing a multitude of geometries each of which only holds for special values of the particle parameters mentioned. But this is obviously geometrization only in a Pickwickian sense.

BIBLIOGRAPHY

Albert, D. and Loewer, B., 'Interpreting the Many Worlds Interpretation', *Synthese* **77** (1988), pp. 195–213.

Ballentine, L. E., 'Can the Statistical Postulate of Quantum Theory Be Derived? A Critique of the Many-Universes Interpretation', *Foundations of Physics* **3** (1973), pp. 229–240.

Balzer, W., Moulines, C. U. and Sneed, J. D., *An Architectonic for Science. The Structuralist Program* (Dordrecht: Reidel, 1987).

Bell, J. S., 'The Measurement Theory of Everett and de Broglie's Pilot Wave', *in* M. Flato *et al.* (eds.), *Quantum Mechanics, Determinism, Causality, and Particles. An International Collection of Contributions in Honor of Louis de Broglie on the Occasion of the Jubilee of His Celebrated Thesis* (Dordrecht: Reidel, 1976, pp. 11–17).

Ben-Menahem, Y., 'Equivalent Descriptions', *The British Journal for the Philosophy of Science* **41** (1990), pp. 261–279.

Berthollet, C. L., *Essai de statique chimique* (Paris: Firmin Didot, 1803).

Bohr, N., 'The Quantum Postulate and the Recent Development of Atomic Theory', *Nature* **121** (1928), Supplement, pp. 580–590.

Bohr, N., 'Kausalität und Komplementarität', *Erkenntnis* **6** (1936), pp. 293–303.

Bohr, N., 'Discussion with Einstein on Epistemological Problems in Atomic Physics', *in* Schilpp 1949, pp. 199–241.

Bohr, N., 'Quantum Physics and Philosophy. Causality and Complementarity', *in* R. Klibansky (ed.), *Philosophy in the Mid-Century I. Logic and Philosophy of Science* (Florence: La nuova italia editrice, 1961, pp. 308–314).

Bonevac, D. A., *Reduction in the Abstract Sciences* (Indianapolis: Hackett, 1982).

Boyle, R., *The Sceptical Chymist*, *in* T. Birch (ed.), *The Works of the Honourable Robert Boyle I* (London 1772, repr.: Hildesheim: Olms, 1965, pp. 458–586).

Bridgman, P. W., *The Logic of Modern Physics* (New York: MacMillan, 1961).

Bridgman, P. W., *A Sophisticate's Primer on Relativity* (1961) (Middletown Conn.: Wesleyan University Press, ^{2}1983).

Brush, S. G., *The Kind of Motion We Call Heat. A History of the Kinetic Theory of Gases in the 19th Century II* (Amsterdam: North-Holland Publishing Co., 1976).

Cannizzaro, S., *Abriss eines Lehrganges der theoretischen Chemie*, ed. L. Meyer, trans. A. Miolati (Leipzig: Engelmann, 1891).

Carnap, R., *Der Raum. Ein Beitrag zur Wissenschaftslehre* (*Kant-Studien Ergänzungshefte 56*) (Berlin: Reuther & Richard, 1922).

Carnap, R., 'Pseudoproblems in Philosophy' (1928), *in* Carnap, *The Logical Structure of the World and Pseudoproblems in Philosophy*, trans. R. A. George (Berkeley: University of California Press, 1967, pp. 301–343).

Carnap, R., 'Wahrheit und Bewährung' (1936), *in* G. Skirbekk (ed.), *Wahrheitstheorien. Eine Auswahl aus den Diskussionen über Wahrheit im 20. Jahrhundert* (Suhrkamp: Frankfurt, 1977, pp. 89–95).

Carnap, R., 'Testability and Meaning', *Philosophy of Science* **3** (1936), pp. 419–471; **4** (1937), pp. 1–40.

Carnap, R., 'The Methodological Character of Scientific Concepts', *in* Feigl/Scriven 1956, pp. 38–76.

Carrier, M., 'Die begriffliche Entwicklung der Affinitätstheorie im 18. Jahrhundert. Newtons Traum – und was daraus wurde', *Archive for History of Exact Sciences* **36** (1986) (a), pp. 327–389.

Carrier, M., 'Zum korpuskularen Aufbau der Materie bei Stahl und Newton', *Sudhoffs Archiv* **70** (1986) (b), pp. 1–17.

Carrier, M., 'Circles without Circularity. Testing Theories by Theory-laden Observations', *in* J. R. Brown and J. Mittelstrass (eds.), *An Intimate Relation. Studies in the History and Philosophy of Science. Presented to Robert E. Butts on his 60th Birthday* (Dordrecht: Kluwer Academic Publishers, 1989, pp. 405–428).

Carrier, M., 'Constructing or Completing Physical Geometry? On the Relation between Theory and Evidence in Accounts of Space-Time Structure', *Philosophy of Science*, **57** (1990), pp. 369–394.

Carrier, M., 'On the Disunity of Science or Why Psychology is not a Branch of Physics', *in* Akademie der Wissenschaften zu Berlin (ed.), *Einheit der Wissenschaften* (Berlin: de Gruyter, 1991, pp. 39–59).

Carrier, M., 'What is Wrong with the Miracle Argument?', *Studies in History and Philosophy of Science* **22** (1991), pp. 23–36.

Carrier, M., 'Physical Force or Geometrical Curvature? Einstein, Grünbaum, and the Measurability of Physical Geometry', *in* J. Earman *et al.* (eds.), *Philosophical Problems of the Internal and External Worlds. Essays on the Philosophy of Adolf Grünbaum* (Pittsburgh: University of Pittsburgh Press, Konstanz: Universitätsverlag, 1993).

Carrier M. and Mittelstrass, J., 'The Unity of Science,' *International Studies in the Philosophy of Science* **4** (1990), pp. 17–31.

Carrier M. and Mittelstrass, J., *Mind, Brain, Behavior. The Mind-Body Problem and the Philosophy of Psychology* (Berlin: de Gruyter, 1991).

Clifton, R. K., 'Some Recent Controversy over the Possibility of Experimentally Determining Isotropy in the Speed of Light', *Philosophy of Science* **56** (1989), 688–696.

Cronbach L. J. and Meehl, P. E.,'Construct Validity in Psychological Tests', *in* Feigl/Scriven 1956, pp. 174–204.

Dalton, J., *A New System of Chemical Philosophy I* (London: Bickerstaff, 1808), (repr. London 1953).

Dancy, J., *An Introduction to Contemporary Epistemology* (Oxford: Basil Blackwell, 1985).

Derksen, A. A., 'Does the Tally Argument Make Freud a Sophisticated Methodologist?', *Philosophy of Science* **59** (1992), pp. 75–101.

Deser, S., 'Self-Interaction and Gauge Invariance', *General Relativity and Gravitation* **1** (1970), pp. 9–18.

Deser, S., Nieuwenhuizen P. van and Boulware, D., 'Uniqueness and Nonrenormalizability of Quantum Gravitation', *in* G. Shaviv and J. Rosen (eds.), *General Relativity and Gravitation* (New York: Wiley, Jerusalem: Israel Universities Press, pp. 1–18).

DeWitt, B. S., 'Quantum Mechanics and Reality', *Physics Today*, **23** (September 1970), pp. 30–35.

Diederich, W., *Strukturalistische Rekonstruktionen* (Braunschweig: Vieweg, 1981).

Dieks, D., 'Gravitation as a Universal Force', *Synthese* **73** (1987), pp. 381–397.

Dieks, D., 'On some Alleged Difficulties in the Interpretation of Quantum Mechanics', *Synthese* **86** (1991), pp. 77–86.

Duhem, P., *The Aim and Structure of Physical Theory* (1906), trans. P. P. Wiener (New York: Atheneum, 1974).

Earman J. (ed.), *Testing Scientific Theories* (*Minnesota Studies in the Philosophy of Science X*), (Minneapolis: University of Minnesota Press, 1983).

Earman J. and Friedman, M., 'The Meaning and Status of Newton's Law of Inertia and the Nature of Gravitational Forces', *Philosophy of Science* **40** (1973), pp. 329–359.

Earman, J., Glymour C. and Stachel J. (eds.), *Foundations of Space-Time Theories* (*Minnesota Studies in the Philosophy of Science VIII*) (Minneapolis: University of Minnesota Press, 1977).

Eder, G., *Elektrodynamik* (Mannheim: Bibliographisches Institut, 1967).

Edidin, A., 'Bootstrapping without Bootstraps', *in* Earman (1983), pp. 43–54.

Edidin, A., 'From Relative Confirmation to Real Confirmation', *Philosophy of Science*, **56** (1988), pp. 265–271.

Ehlers, J., 'The Nature and Structure of Spacetime', *in* J. Mehra (ed.), *The Physicist's Conception of Nature* (Dordrecht: Reidel, 1973 (a), pp. 71–91).

Ehlers, J., 'Survey of General Relativity Theory', *in* W. Israel (ed.), *Relativity, Astrophysics and Cosmology* (Dordrecht: Reidel, 1973 (b), pp. 1–125).

Ehlers, J., 'Einführung in die Raum-Zeit-Struktur mittels Lichtstrahlen und Teilchen', *in* J. Audretsch and K. Mainzer (eds.), *Philosophie und Physik der Raum-Zeit* (Mannheim: Bibliographisches Institut, 1988, pp. 145–162).

Ehlers J. and Köhler, E., 'Path Structures on Manifolds', *Journal of Mathematical Physics* **18** (1977), pp. 2014–2018.

Ehlers, J., Pirani F. A. E. and Schild, A., 'The Geometry of Free Fall and Light Propagation', *in* L. O'Raifeartaigh (ed.), *General Relativity. Papers in Honour of J. L. Synge* (Oxford: Clarendon, 1972, pp. 63–84).

Einstein, A., 'On the Electrodynamics of Moving Bodies' (1905), *in* Lorentz *et al.* (1923), pp. 35–65.

Einstein, A., 'The Foundations of the General Theory of Relativity' (1916), *in* Lorentz *et al.* (1923), pp. 109–164.

Einstein, A., *Über die spezielle und die allgemeine Relativitätstheorie* (1917) (Braunschweig: Vieweg, [21]1969), (repr. 1985).

Einstein, A., 'Prinzipielles zur allgemeinen Relativitätstheorie', *Annalen der Physik* **55**

(1918), pp. 241–244.
Einstein, A., 'Remark', *Physikalische Zeitschrift* **21** (1920), p. 662.
Einstein, A., 'Geometrie und Erfahrung' (1921), *in* C. Seelig (ed.), *Albert Einstein. Mein Weltbild* (Frankfurt: Ullstein, 1984, pp. 119–127).
Einstein, A., *The Meaning of Relativity* (1922) (Princeton: Princeton University Press, ^{5}1955).
Einstein, A., 'Autobiographical Notes/Remarks on the Essays Appearing in this Collective Volume', *in* Schilpp 1949, pp. 1–94, pp. 663–688.
Ellis B. and Bowman, P., 'Conventionality in Distant Simultaneity', *Philosophy of Science* **34** (1967), pp. 116–136.
Feigl, H., 'Existential Hypotheses. Realistic versus Phenomenalistic Interpretations', *Philosophy of Science* **17** (1950), pp. 35–62.
Feigl H. and Scriven M. (eds.), *The Foundations of Science and the Concepts of Psychology and Psychoanalysis* (*Minnesota Studies in the Philosophy of Science I*) (Minneapolis: University of Minnesota Press, 1956).
Freud, S., *The Interpretation of Dreams* (1900), ed./trans. J. Strachey, *The Standard Edition of the Complete Psychological Works of Sigmund Freud IV–V* (London: Hogarth, 1953, pp. 1–621).
Freud, S., *The Psychopathology of Everyday Life* (1901), ed./trans. J. Strachey, *The Standard Edition of the Complete Psychological Works of Sigmund Freud VI* (London: Hogarth, 1960).
Freud, S., *An Outline of Psycho-Analysis* (1940), ed./trans. J. Strachey, *The Standard Edition of the Complete Psychological Works of Sigmund Freud XXIII* (London: Hogarth, 1964, pp. 144–207).
Friedman, M., 'Simultaneity in Newtonian Mechanics and Special Relativity', *in* Earman, Glymour and Stachel (1977), pp. 403–432.
Friedman, M., *Foundations of Space-Time Theories. Relativistic Physics and Philosophy of Science* (Princeton: Princeton University Press, 1983).
Glymour, C., 'The Epistemology of Geometry', *Nous* **11** (1977), pp. 227–251.
Glymour, C., *Theory and Evidence* (Princeton: Princeton University Press, 1980).
Glymour, C., 'On Testing and Evidence', *in* Earman (1983), pp. 3–26.
Grandy, R. E., 'Some Comments on Confirmation and Selective Confirmation', *Philosophical Studies* **18** (1967), pp. 19–24.
Grimes, T. R., 'Truth, Content, and the Hypothetico-Deductive Method', *Philosophy of Science* **57** (1990), pp. 514–522.
Grünbaum, A., 'The Duhemian Argument', *Philosophy of Science* **27** (1960), pp. 75–87.
Grünbaum, A., *Philosophical Problems of Space and Time* (Dordrecht: Reidel, ^{2}1973).
Grünbaum, A., 'Retrospective vs. Prospective Testing of Aetiological Hypotheses in Freudian Theory', *in* Earman (1983), pp. 315–347.
Grünbaum, A., *The Foundations of Psychoanalysis. A Philosophical Critique* (Berkeley: University of California Press, 1984).
Grünbaum A. and Janis, A. I., 'The Geometry of the Rotating Disk in the Special Theory of Relativity', *in* W. C. Salmon (ed.), *Hans Reichenbach. Logical Empiricist*

(Dordrecht: Reidel, 1979, pp. 321–339).

Hanson, N. R., *Patterns of Discovery. An Inquiry into the Conceptual Foundations of Science* (1958) (Cambridge: Cambridge University Press, 1965).

Havas, P., 'Simultaneity, Conventionalism, General Covariance, and the Special Theory of Relativity', *General Relativity and Gravitation* **19** (1987), pp. 435–453.

Heilig U. and Pfister, H., 'Characterization of Free Fall Paths by a Global or Local Desargues Property', *Journal of Geometry and Physics* **7** (1990), pp. 419–446.

Helmholtz, H. v., 'Über die Thatsachen, die der Geometrie zum Grunde liegen' (1868), *in* Helmholtz, *Über Geometrie* (Darmstadt: Wissenschaftliche Buchgesellschaft, 1968, pp. 32–60).

Helmholtz, H. v., 'Über den Ursprung und die Bedeutung der geometrischen Axiome' (1870), *in* Helmholtz, *Über Geometrie* (Darmstadt: Wissenschaftliche Buchgesellschaft, 1968, pp. 1–31).

Hempel, C. G., 'Studies in the Logic of Confirmation' (1945), *in* Hempel (1965), pp. 3–46.

Hempel, C. G., *Fundamentals of Concept Formation in Empirical Science* (1952), (*International Encyclopedia of Unified Science II.7*) (Chicago: University of Chicago Press, ^{7}1964).

Hempel, C. G., 'The Theoretician's Dilemma: A Study in the Logic of Theory Construction' (1958), *in* Hempel (1965), pp. 173–226.

Hempel, C. G., *Aspects of Scientific Explanation And Other Essays in the Philosophy of Science* (New York: Free Press, London: Collier-MacMillan, 1965).

Hempel, C. G., *Philosophy of Natural Science* (Englewood Cliffs N. J.: Prentice Hall, 1966).

Herbert, N., *Quantum Reality. Beyond the New Physics* (New York: Doubleday, 1985).

Hoyningen-Huene, P., *Die Wissenschaftsphilosophie Thomas S. Kuhns. Rekonstruktion und Grundlagenprobleme* (Braunschweig: Vieweg, 1989).

Hoyningen-Huene, P., 'Kuhn's Conception of Incommensurability', *Studies in History and Philosophy of Science* **21** (1990), pp. 481–492.

Jammer, M., *The Philosophy of Quantum Mechanics. The Interpretations of Quantum Mechanics in Historical Perspective* (New York: Wiley, 1974).

Kamlah, A., 'Kepler im Lichte der modernen Wissenschaftstheorie', *in* H. Lenk (ed.), *Neue Aspekte der Wissenschaftstheorie* (Braunschweig: Vieweg, 1971, pp. 205–220).

Kanitscheider, B., *Geometrie und Wirklichkeit* (Berlin: Duncker & Humblot, 1971).

Kant, I., *Metaphysical Foundations of Natural Science* (1786), trans. J. W. Ellington, *in Immanuel Kant. Philosophy of Material Nature* (Indianapolis: Hackett, 1985, pp. 1–134).

Karsten, D. L. G., 'Bemerkungen über Berthollets chemische Affinitätslehre', *Allgemeines Journal der Chemie* **10** (1803), pp. 135–156.

Kuhn, T. S., *The Structure of Scientific Revolutions*, (*International Encyclopedia of Unified Science II.2*) (Chicago: University of Chicago Press, 1962).

Kuhn, T. S., 'Objectivity, Value Judgment, and Theory Choice', *in* Kuhn, *The Essential Tension* (Chicago: University of Chicago Press, 1977, pp. 320–339).

Kuhn, T. S., 'Commensurability, Comparability, Communicability', *in* P. D. Asquith and T. Nickles (eds.), *PSA 1982 II* (East Lansing Mich.: Philosophy of Science Association, 1983, pp. 669–688).

Kretschmann, E., 'Über den physikalischen Sinn der Relativitätspostulate. A. Einsteins neue und seine ursprüngliche Relativitätstheorie', *Annalen der Physik* **53** (1917), pp. 575–614.

Lambert K. and Brittan, G. G., *An Introduction to the Philosophy of Science* (Atascadero Cal.: Ridgeview Publishing Co., [3]1987).

Landau L. D. and Lifshitz, E. M., *Lehrbuch der theoretischen Physik II. Klassische Feldtheorie* (1976) (Berlin: Akademie-Verlag, [11]1989).

Lakatos, I., 'Falsification and the Methodology of Scientific Research Programmes' (1970), *in* J. Worrall and G. Currie (eds.), *The Methodology of Scientific Reserach Programmes* (*Philosophical Papers I*) (Cambridge: Cambridge University Press, 1978, pp. 8–101).

Laudan, L., *Progress and its Problems. Towards a Theory of Scientific Growth* (Berkeley, University of California Press, 1977).

Lienert, G. A., *Testaufbau und Testanalyse* (Weinheim: Beltz, [3]1969).

Lorentz H. A. *et al.*, *The Principle of Relativity. A Collection of Original Memoirs on the Special and General Theory of Relativity* (1923), trans. W. Perrett and G. B. Jeffery, (New York: Dover, 1952).

Mainzer, K., *Geschichte der Geometrie* (Mannheim: Bibliographisches Institut, 1980).

Malament, D., 'Causal Theories of Time and the Conventionality of Simultaneity', *Nous* **11** (1977), pp. 293–300.

Marzke R. F. and Wheeler, J. A., 'Gravitation as Geometry I: The Geometry of Space-Time and the Geometrodynamical Standard Meter', *in* H.-Y. Chiu and W. F. Hoffmann (eds.), *Gravitation and Relativity* (New York: Benjamin, 1964, pp. 40–64).

Mayo, D. G., 'Novel Evidence and Severe Tests', *Philosophy of Science* **58** (1991), pp. 523–552.

Meehl, P. E., 'Subjectivity in Psychoanalytic Inference: The Nagging Persistence of Wilhelm Fliess's Achensee Question', *in* Earman (1983), pp. 349–411.

Misner, C. W., Thorne K. S. and Wheeler, J. A., *Gravitation* (San Francisco: Freeman, 1973).

Nagel, E., *The Structure of Science. Problems in the Logic of Scientific Explanation* (London: Routledge & Kegan Paul, 1961).

Nash, L. K., *The Atomic-Molecular Theory* (*Harvard Case Studies in Experimental Science IV*) (Cambridge Mass.: Harvard University Press, 1967).

Newton, I., *Mathematical Principles of Natural Philosophy*, trans. A. Motte and F. Cajori, [3]1726, 1729, (New York: Greenwood, 1969).

Norton, J., 'What was Einstein's Principle of Equivalence?', *Studies in History and Philosophy of Science* **16** (1985), pp. 203–246.

Norton, J., 'Coordinates and Covariance: Einstein's View of Space-Time and the Modern View', *Foundations of Physics* **19** (1989), pp. 1215–1263.

Ohanian, H. C., 'What is the Principle of Equivalence?', *American Journal of Physics*

45 (1977), pp. 903–909.
Papineau, D., *Theory and Meaning* (Oxford: Clarendon, 1979).
Perlick, V., 'Characterization of Standard Clocks by Means of Light Rays and Freely Falling Particles', *General Relativity and Gravitation* **19** (1987), 1059–1073.
Poincaré, H., 'La mesure du temps', *Revue de métaphysique et de morale* **6** (1898), pp. 1–13.
Poincaré, H., *La science et l'hypothèse* (1902) (Paris: Flammarion, 1968).
Popper, K. R., *The Logic of Scientific Discovery* (1934) (London: Hutchinson, 1959).
Putnam, H., 'The Analytic and the Synthetic' (1962), *in* Putnam, *Mind, Language and Reality*, (*Philosophical Papers II*) (Cambridge: Cambridge University Press, 1975, pp. 33–69).
Putnam, H., 'An Examination of Grünbaum's Philosophy of Geometry' (1963), *in* Putnam, *Mathematics, Matter and Method*, (*Philosophical Papers I*) (Cambridge: Cambridge University Press, 1975, pp. 93–129).
Putnam, H., 'The Refutation of Conventionalism', *in* Putnam, *Mind, Language and Reality*, (*Philosophical Papers II*) (Cambridge: Cambridge University Press, 1975, pp. 153–191).
Putnam, H., 'Equivalence', *in* Putnam, *Realism and Reason*, (*Philosophical Papers III*) (Cambridge: Cambridge University Press, 1983, pp. 26–45).
Rae, A., *Quantum Physics: Illusion or Reality?* (Cambridge: Cambridge University Press, 1986).
Ray, C., *The Evolution of Relativity* (Bristol: Adam Hilger, 1987).
Redhead, M., *Incompleteness, Nonlocality, and Realism. A Prolegomenon to the Philosophy of Quantum Mechanics* (Oxford: Clarendon Press, 1987).
Reichenbach, H., *Axiomatik der relativistischen Raum-Zeit-Lehre* (1924) (Braunschweig: Vieweg, 1965).
Reichenbach, H., *The Philosophy of Space and Time* (1928) (New York: Dover, 1965).
Reichenbach, H., 'The Principle of Causality and the Possibility of its Empirical Confirmation' (1932), *in* Reichenbach, *Modern Philosophy of Science. Selected Essays*, ed./trans. M. Reichenbach, (London: Routledge & Kegan Paul, 1959, pp. 109–134).
Reichenbach, H., *Experience and Prediction. An Analysis of the Foundations and the Structure of Knowledge* (1938) (Chicago: University of Chicago Press, [6]1966).
Reichenbach, H., *The Rise of Scientific Philosophy* (1951) (Berkeley: University of California Press, [11]1964).
Richardson, R. C., 'The 'Tally-Argument' and the Validation of Psychoanalysis', *Philosophy of Science* **57** (1990), pp. 668–676.
Salmon, W. C., 'The Conventionality of Simultaneity', *Philosophy of Science* **36** (1969), pp. 44–63.
Salmon, W. C., 'The Philosophical Significance of the One-Way Speed of Light', *Nous* **11** (1977), pp. 252–292.
Schilpp P. A. (ed.), *Albert Einstein. Philosopher-Scientist* (1949) (La Salle Ill.: Open Court, [3]1970).
Schlick, M., 'Meaning and Verification' (1936), *in* H. Schleichert (ed.), *Logischer*

Empirismus. Der Wiener Kreis (Munich: Wilhelm Fink, 1975, pp. 118–147).

Schrödinger, E., 'The Present Situation in Quantum Mechanics' (1935), trans. J. D. Trimmer, *in* J. A. Wheeler and W. H. Zurek (eds.), *Quantum Theory and Measurement* (Princeton: Princeton University Press, 1983, pp. 152–167).

Scott J. F. (ed.), *The Correspondence of Isaac Newton IV* (Cambridge: Cambridge University Press, 1967).

Sexl, R. U., 'Theories of Gravitation', *Fortschritte der Physik* **15** (1967), pp. 269–307.

Sexl R. U. and Urbantke, H. K., *Gravitation und Kosmologie. Eine Einführung in die Allgemeine Relativitätstheorie* (Mannheim: Bibliographisches Institut, 1983).

Sklar, L., *Space, Time, and Spacetime* (Berkeley: University of California Press, 1974).

Sklar, L., 'Facts, Conventions, and Assumptions in the Theory of Space-Time', *in* Earman, Glymour and Stachel (1977), pp. 206–274.

Sneed, J. D., *The Logical Structure of Mathematical Physics* (Dordrecht: Reidel, [2]1979).

Stegmüller, W., *The Structuralist View of Theories. A Possible Analogue of the Bourbaki Programme in Physical Science* (Berlin: Springer, 1979).

Stegmüller, W., *Theorie und Erfahrung III. Die Entwicklung des neuen Strukturalismus seit 1973* (Berlin: Springer, 1986).

H. Stephani, *General Relativity. An Introduction to the Theory of the Gravitational Field* (1980) (Cambridge: Cambridge University Press, 1982).

Suen H. K. and Ary, D., *Analyzing Quantitative Behavioral Data* (Hillsdale N.J.: Lawrence Erlbaum, 1989).

Synge, J., *Relativity: The General Theory* (1960) (Amsterdam: North-Holland, [4]1971).

Thirring, W., 'Lorentz-invariante Gravitationstheorien', *Fortschritte der Physik* **7** (1959), pp. 79–101.

Thirring, W., 'An Alternative Approach to the Theory of Gravitation', *Annals of Physics* **16** (1961), pp. 96–117.

Wald, R. M., *General Relativity* (Chicago: University of Chicago Press, 1984).

Weinberg, S., *Gravitation and Cosmology. Principles and Applications of the General Theory of Relativity* (New York: Wiley, 1972).

Weyl, H., *Raum, Zeit, Materie. Vorlesungen über allgemeine Relativitätstheorie* (1923a) (Darmstadt: Wissenschaftliche Buchgesellschaft, [6]1961).

Weyl, H., 'Mathematische Analyse des Raumproblems' (1923b), *in* Weyl, *Mathematische Analyse des Raumproblems / Was ist Materie?* (Darmstadt: Wissenschaftliche Buchgesellschaft, 1977, pp. 1–117).

Weyl, H., *Philosophy of Mathematics and Natural Science* (Princeton: Princeton University Press, 1949).

Winnie, J. A., 'Special Relativity without One-Way Velocity Assumptions', *Philosophy of Science* **37** (1970) pp. 81–99, pp. 223–238.

Winnie, J. A., 'The Causal Theory of Space-Time', *in* Earman, Glymour and Stachel (1977), pp. 134–205.

Wyss, W., 'Zur Unizität der Gravitationstheorie', *Helvetica Physica Acta* **38** (1965), pp. 469–480.

INDEX OF NAMES

INDEX OF SUBJECTS